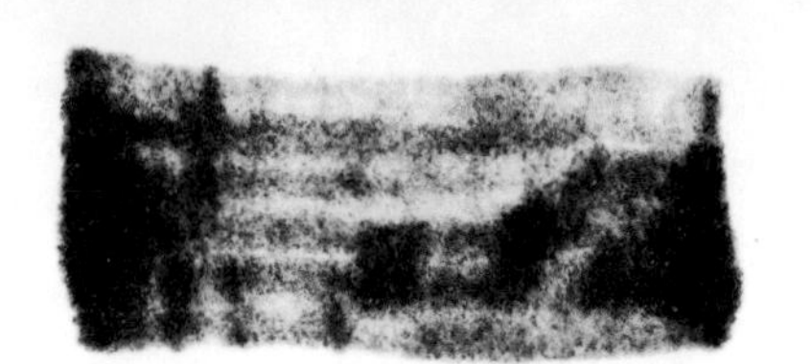

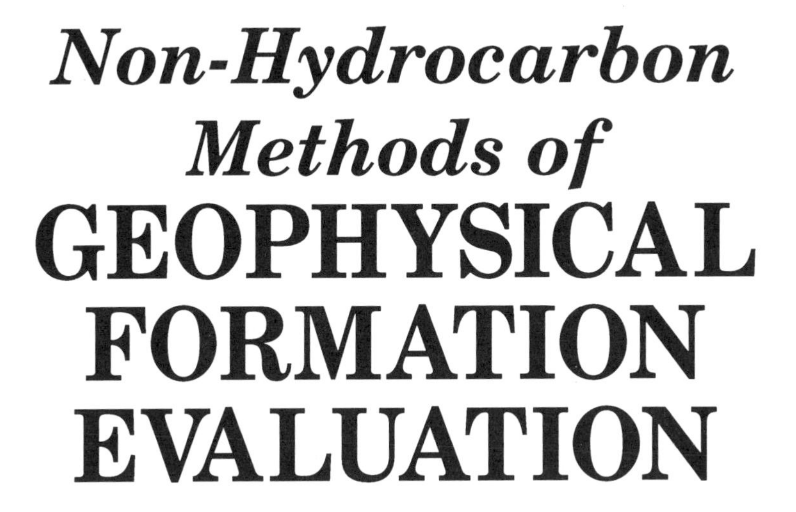

Non-Hydrocarbon Methods of GEOPHYSICAL FORMATION EVALUATION

James K. Hallenburg

Lewis Publishers

Boca Raton Boston New York Washington London

Library of Congress Cataloging-in-Publication Data

Hallenburg, James K.
Nonhydrocarbon methods of geophysical formation evaluation / James K. Hallenburg.
p. cm.
Includes bibliographical references and index.
ISBN 1-56670-262-3 (alk. paper)
1. Engineering geology—Instruments. 2. Geophysical instruments. 3. Geophysics in archaeology. 4. prospecting—Geophysical methods. I. Title.
TA705.H34 1997
551′.028—dc21

97-29081
CIP

Lewis Publishers is an imprint of CRC Press LLC

International Standard Book Number 1-56670-262-3
Library of Congress Card Number 97-29081
Printed in the United States of America 1 2 3 4 5 6 7 8 9 0
Printed on acid-free paper

Preface

Electrical methods are the oldest of the commercial geophysical methods of gathering information for formation evaluation. It became obvious in the late 19th century that traditional surface geological exploration needed methods of examining deeper than the surface and shallow depths into the earth that were then in use. They were also needed on a regular basis. Thus, temperature, natural potentials, gravity, and electrical resistivity methods were investigated and later developed into viable commercial tools.

Later, in the 20th century, the existing surface methods were adapted to downhole use. The Schlumberger brothers, Conrad and Marcel, for example, had an active operation in the mines in Europe, using electromagnetic methods. They then applied the surface dipole resistivity arrays to downhole use in the coal fields at Pechelbron in France. The original resistivity system used by Conrad Schlumberger was similar to a surface dipole-dipole. He simply adapted the existing surface method to downhole use. A little bit of the history of logging geophysics is presented in *Introduction to Geophysical Formation Evaluation*, a companion text to this volume.

This was just at the time that the demand for petroleum was increasing dramatically because of the widespread and rapidly growing use of the internal combustion engine. The new petroleum industry was looking for better exploration and development tools. They found a tremendous one in geophysics. Large efforts of time and money went into the expansion of geophysical methods, especially the downhole methods.

Thus, the vast bulk of examples, papers, books, discussions, teaching effort, and financing went into the use of geophysics **for the search for OIL**. This has lasted for more than 70 years.

About 1960, the mineral industry discovered that geophysics was useful in the search for uranium, then coal, and then water. From this, the use has spread to environmental uses, oceanography, and many other disciplines, including real estate and law. One of the latest has been archeology. The total budget for all of these, however, is still **minor** compared to the current budget for oilfield geophysics.

The upshot of this situation is that we do not have the organizations (they are beginning to appear but do not have the support they deserve), the papers, the examples, the textbooks, the college-level classes, or the research effort in nonpetroleum that the petroleum industry has generated over 80 or more years.

Meanwhile, we can still borrow from the petroleum industry. We must remember, however, that there are fundamental differences in scope, end use, philosophy, design, and economics between the petroleum geophysics and nonpetroleum geophysics. We are looking for different things, with tools that are different (or should be). This last comment is even reflected within the title of this book — *Non-Hydrocarbon Geophysics.*

Development of downhole geophysical devices followed three paths: (1) interpretation of the results (the measurements), (2) control of the size of the volume of investigation, and (3) the development of newer systems. The basic ideas of most of these are covered in *Introduction to Geophysical Formation Evaluation.* The standard methods are covered in greater detail in *Standard Methods of Geophysical Formation Evaluation,* another companion text to this volume.

In all of the early systems the depth of investigation of the measurement and the resolution were coupled. If the depth of investigation (horizontal in the case of downhole methods and vertical in the case of surface methods and horizontal holes) was increased, the resolution was proportionally decreased. In addition, distortion of the curve shape or data was frequently a problem. The normal resistivity device (the surface dipole), for example, always has distortion at the bed boundaries. The curve does not show the expected sinusoidal response but has a curve, a flat space, and a continuation of the curve across the boundary. The lateral device is even worse. Because it is the difference between two normal devices, the response curve is unsymmetrical and depends upon the bed thickness. Data reduction of the measurements made with these devices and, especially interpretation, is more of an art form than a science.

The results of efforts to change/improve/control these problems have been whole new systems for improved general and specialized uses. While the standard systems still persist in the records and in use, they are being replaced by newer systems which will give less distorted, more easily interpreted information.

Much effort has been expended, also, in developing suitable combinations of methods for specific purposes. This has included new and more explicit presentations and better data handling. A further improvement involves making automatic and real-time environmental corrections. These newer systems usually depend upon the recently developed digital data transmission and handling techniques.

These are the systems we will explore in *Non-Hydrocarbon Methods of Geophysical Formation Evaluation.* Along with this we will examine some of the speciality methods and some of the less often used methods. We will also take a look at some of the data handling methods, new and old.

Introduction to Geophysical Formation Evaluation and *Standard Methods of Geophysical Formation Evaluation* concentrate on methods which have been chiefly developed and used by the petroleum industry. There are good reasons for covering these methods and using examples from the petroleum industry. They were developed and paid for by the petroleum industry but can and are used in non-hydrocarbon applications, sometimes without modification. In addition, some of the non-hydrocarbon applications (hydrological usage, for example) are identical to many of the petroleum applications.

Non-Hydrocarbon Methods of Geophysical Formation Evaluation has been deliberately oriented to non-hydrocarbon applications. This book also goes into methods other than downhole geophysical logging.

I would point out that there is no way that these three books could ever hope to cover the geophysical field completely. In the first place, the scope of the field is tremendous. Secondly, the technical aspects of geophysics are changing rapidly but relatively quietly. This is due, in large part, to the rapid changes in digital and (specifically) computer technology. Thus, I have just touched the surface of geophysics for non-hydrocarbon uses. I strongly recommend that a student of this discipline carefully examine the literature of the several organizations dedicated to these ideas. They are the Minerals and Geotechnical Engineering Society (a subchapter of the Society of Professional Well Log Analysis, Houston, Texas), the Environmental and Engineering Geophysical Society, Wheat Ridge, Colorado; and the Society of Exploration Geophysicists, Tulsa, Oklahoma. Their publications are valuable.

Dipmeter Methods

Dipmeter methods have been fully covered in oilfield literature and texts, especially from the major logging contractors. These are referenced in the bibliography of this volume (see Bigelow, 1987, and Schlumberger, 1983).

The non-hydrocarbon use of the dipmeter and its analyses for non-hydrocarbon usage closely follow those for petroleum use. Therefore, this will not be covered further in this volume. Note, however, that the dipmeter methods can be a valuable tool for engineering, geology, and geophysical applications. Especially, the dipmeter should be considered for projects in engineering and commercial hydrology, and also for any other non-hydrocarbon problems where rock mechanics is important. There are suitable dipmeter systems available from some of the non-hydrocarbon logging contractors (i.e., BPB Instruments). The major oilfield logging contractors have very sophisticated dipmeter tools and programs. It must be remembered, however, that most oilfield dipmeter tools have been designed for large diameter boreholes (6 in. [15 cm] or more).

Consider the use of the dipmeter techniques wherever dip, stratification, or lamination are unknown, obscure, and important. The dipmeter can also be used as a fracture locator, especially the four-curve dipmeter.

Others

There are many more available geophysical methods which are not discussed in *Non-Hydrocarbon Methods of Geophysical Formation Evaluation.* These include some of the newer methods and presentations which appear to have little use outside of the petroleum industry. And, of course, this author cannot be fully conversant with all methods, as this field is expanding rapidly.

The Bibliography and References section of the Appendices of *Non-Hydrocarbon Methods of Geophysical Formation Evaluation* has deliberately been made quite large, as many of the items are obscure and hard to trace.

I have a large file (albeit, still quite incomplete) of papers and brochures collected over a period of about 50 years and have drawn on this extensively.

A Word of Warning!!

Numerous tables have been included in the following text. The values listed are subject to change. These are being upgraded almost daily. Use the values for estimation **only**. If you have a critical calculation or determination to make, the value listed in this text may be correct, **but** be sure to find the best known value and verify the one from this text.

James K. Hallenburg
Author

The Author

James K. Hallenburg is a geophysicist, petrophysicist in Tulsa, Oklahoma. He is a graduate of Northwestern University and has attended several other universities and colleges. He was a pilot in the U.S. Army Air Corps and returned to finish his education. He designed geophysical systems for Schlumberger Well Services, Inc. for 18 years. He was a Senior Engineer for the Mohole Project and Chief Engineer for the Western Company of North America. He later operated Data Line Logging Company out of Casper, Wyoming and then was Manager of Applications Engineering for Century Geophysical in Tulsa. He was a consultant for the International Atomic Energy Agency. He is the author, co-author, and editor of several books and published computer programs. He gave seminars in numerous countries and taught at Casper College and the University of Tulsa Petroleum Engineering Department.

Introduction to Non-Geophysical Hydrocarbon Methods

Introduction

The use of geophysics has been in existence for many centuries. In fact, some form of it has been used ever since man started to look for minerals for paint and decoration, stone for tools, metal ore for tools and decoration, glasses for decoration and glazing, and many other purposes. Some of the methods were effective and are still in use. Others have been superseded by more modern methods and instrumentation. Still others were found to be incorrect or inaccurate and were discarded (a process that still goes on). Some methods were myths that persist today.

Geophysics is a tool. It should be used, if it is warranted. It should not be used simply for the sake of using it. It is a powerful and often expensive tool that should be used with planning and common sense.

The majority of modern geophysical methods were adapted to petroleum exploration and have been highly successful. This success has resulted in the oil industry investing large sums in developing and refining the methods, and the result has been tremendous. Geophysics has almost become synonymous with the search for oil.

This volume is intended to point out that geophysics can be used profitably for purposes other than oil exploration and development. Non-hydrocarbon efforts owe a lot to the oil industry. It should be remembered, however, that the oil people are searching for a material different from the mineral people, or the archeologist, or the environmentalist, or the civil engineer. Oil people are looking for the fluids which fill the pore spaces in sediments. A miner is only interested in these fluids as they interfere with his efforts. He is looking for a solid rock-like material. The archeologist is looking for a buried building or grave site. These are very different from a hydrocarbon deposit.

Some Warnings

Individual measurements should, of course, never be used alone. There are several good reasons for this caution:

1. The most obvious is that if two or more separate measurements give the same or nearly the same results, confidence in the value is increased tremendously.
2. Remember that we are working in an environment which, especially at first, is largely unknown. If any information exists, it is probably in the sketchiest of terms.
3. The working environment is usually quite complex.
4. We are not always sure of the extent or volume of the measurement. This is something which easily can change drastically with the type of parameter and the type of measurement.
5. Most of the time the desired parameter is only measured indirectly. The end result must be calculated or inferred.

This is a normal situation in engineering, regardless of the tool being used.

It is wise to always repeat measurements, if possible. Of course, it is not always possible or practicable. But, cost is a poor decider of priorities in this case. Always use more than one type of measurement to determine any vital parameter. This relates back to the first idea in the previous paragraph. For example, if determining porosity from a borehole, use a density and a neutron measurement or use a density measurement and a laboratory determination from a core sample.

Always measure a bit more than your specifications call for. If, for instance, you are interested in the bed rock down to 100 ft (30 m), drill down to 125 ft (38 m) or 150 ft (46 m) and measure from the bottom up. Borrowing from oilfield experience, a number of good horizons have been discovered below the target sand.

The cheapest contractor may be the costliest. Always go with the **best** contractor. This will invariably be the least costly. Poor information is worse than useless, it often is misleading because we do not know that it is erroneous. In the long run, the best is usually the least expensive.

One should consult a geophysicist before any new and extensive project. The geophysicist should consult with the project geologist to plan what is needed to acquire the necessary information in the least costly way. Remember, too, that the logging contractor's engineer is an expert with the equipment. The engineer knows the capabilities of his equipment and how to reduce the data. On the other hand, the major logging contractors are oilfield oriented. Further, they are interested in selling their own services. And, in general their experience is with **downhole** methods. One is better off retaining an independent, non-hydrocarbon geophysicist or geologist for, at least, the beginning of a new project. In all fairness, the major oilfield logging contractors and many of the minors are quite good at their jobs.

Finally, as taught in the U.S. Air Force, trust your instruments, not your hunch or desires. If the proper instruments are chosen and well operated, their readings are good.

Contents

Appendices

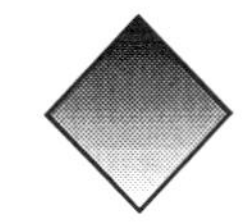

Non-Hydrocarbon Methods of GEOPHYSICAL FORMATION EVALUATION

1

Resistivity Methods

1.1 Discussion about Resistivity in Formation Evaluation

Electrical resistivity (and conductivity) measurements occupy a unique place in petroleum formation evaluation. Of course, the use of resistivity measurements is not an exclusive domain of the petroleum industry. The family of electrical resistivity measurements is an important component of *all* formation evaluation. The good resolution of the single point resistivity or resistance curves is invaluable in the search for sedimentary mineral deposits and water. Other quantitative resistivity measurements serve many of the same functions for nonhydrocarbon exploration as for petroleum. The fact is, of course, that quantitative resistivity measurements are not needed as often in nonhydrocarbon uses. Electrical resistivity measurements have an important potential use in any sedimentary project. Use in massive, sedimentary, and nonsedimentary environments is more limited, but it is still worth considering. We will see later in this text that the resistivity values can be very diagnostic.

Surface resistivity measurements are also important in nonhydrocarbon exploration. This was brought out in the previous volumes of this series. Quantitative surface resistivity measurements are frequently combined with the induced polarization (IP) systems. This is advantageous because the resistivity measurement eases the correlation of the IP map with the subsurface logs and surface seismic surveys. Resistivity is a simple and inexpensive addition to the IP circuitry.

There are several good reasons for considering electrical resistivity and conductivity measurements in nonhydrocarbon endeavors. These methods are among the oldest and best researched of the geophysical methods. The Schlumbergers, for instance, were using surface resistivity (and IP) methods commercially well before investigating borehole uses of these means were used. First, the petroleum-oriented systems exist and are readily available. They are different from nonpetroleum instruments, however, and the differences must be understood before any of the present systems are used. Second, the petroleum methods are frequently quite adaptable to nonpetroleum applications. This has been demonstrated by the presently extensive use of these methods for exploration, development, and production of uranium, water, oceanography, and coal. Third, the development of these instruments and methods was overwhelmingly financed by and for the petroleum industry. Fourth, practically all of the references, examples, and values available for demonstrating the use of downhole and surface methods are from petroleum uses.

Electrical resistivity measurements can define fractures, conductive mineral bodies, the presence of air and gases, fluid salinity, stratification, and many other phenomena that are important to mineral geologists, hydrologists, oceanographers, civil engineers, and countless others. Electrical resistivity methods are not the exclusive province of the petroleum industry, even though the petroleum industry has done the vast majority of work involving resistivity.

Laboratory electrical measurements on solid samples are not commonplace. They are not done routinely. However, most core and university laboratories are able to make resistivity measurements on samples. Fluid resistivity measurements are frequently done in the laboratory. In general, laboratory resistivity (and conductivity) measurements on solid samples can be quite accurate because of the controlled conditions of the measurements. They are, however, performed on relatively small samples. Therefore, the statistical variation of the small sample must be taken into account when comparing these values with surface and borehole measurements.

The vast bulk of resistivity measurements are subsurface trolling (logging) measurements made from a borehole. Virtually all of the recent developments in electrical resistivity determinations have involved borehole devices. These developments have been in the array-type systems (many measurements of different geometries are made simultaneously and assembled into various configurations by computer processing), scanner-type systems (many identical highly detailed measurements made over a designated, fixed area and assembled into a visual, picture-type presentation), and presentations (diverse instrument and geometry responses assembled to solve for specific parameters, such as porosity, rock-type, etc.).

The lateral extent (depth of investigation) and shape (geometry) of the volume of investigation of simple resistivity and conductivity devices depend directly upon the spacings between the respective electrodes. The resolution, however, is an inverse function of the spacings. That is, the greater the spacing, the deeper the depth of investigation and the coarser the resolution of the system. This may not be true with some of the more complex, modern systems.

It is worthwhile to remember that, in general, the larger the sample volume, the lower the resolution. In older, simpler methods, the higher resolution was achieved by reducing the measured volume by shortening the electrode or coil spacing. This limitation has been largely overcome by modern focused, computer-controlled systems. Focusing devices have largely replaced the older, simpler methods because of their much higher degree of control of the geometry of investigation. This, in spite of the usual greater complexity.

Solid-state devices, computer control, and computer data handling have made a tremendous improvement, not only in electrical resistivity measurement, but in *all* geophysical methods.

1.1.1 Terminology

Descriptions of phenomena involving the use of the terms "resistivity" or "R" or "Ωm" can infer also the reciprocal units, "conductivity" or "C" or "S/m" and vice versa, unless otherwise stated. Similarly, the term "resistance" or "r" or "Ω" can infer "conductance" or "c" or "S". You will remember that the conductivity and conductance are the reciprocals of the resistivity and resistance, respectively. While the terms R and C (and r and c) can be interchanged by using the reciprocal, there are some practical differences in their usages. Much of this has been discussed in the introductory volume of this series. It is probably worthwhile, however, to enlarge upon that somewhat.

1.1.2 Temperature Effects

The four values, R, r, C, and c, exhibit predictable temperature responses. The variation may be small or large, depending upon the material being investigated, but it will be present. Thus, it is very important to record and specify the temperature of the material *at the time of measurement*. In our nonhydrocarbon work, the temperature specification is most important when working with fluid samples on the surface. When mud, mud components, formation water samples, and other such surface measurements are made, the temperature can easily cause a resistivity variation of ±200%. Lack of the temperature data makes the resistivity value meaningless.

We do not usually specify downhole temperatures when speaking of formation and other downhole resistivities, because the downhole temperature can often be estimated fairly closely. Sometimes there are temperature data available. Formation temperatures do not normally vary rapidly with time. Also, we usually need the downhole formation fluid resistivity values compared to formation resistivity values measured at the same time. We frequently study a whole set of measurements which were made at approximately the same temperature. In this case, temperature variations are unimportant.

Downhole mud component resistivity values are handled quite differently. Usually, the measurement is made on a surface sample at a temperature quite different from the downhole formation temperature. Thus, we calculate a probable downhole temperature (making some assumptions). The assumptions provide a rough estimate of the borehole fluid temperature. Then we further assume that the borehole temperature and the formation temperature are the same when obviously they are not. The may be close, however, and the greater the time of the measurement after the cessation of circulation, the closer the mud values will be to the formation temperature. This is not an ideal situation, but it is a pragmatic one. The electrical resistance of any sample of material, at any temperature (provided there is no change of state nor a chemical change) is

$$r_t = r_0(1 + \alpha T) \tag{1.1}$$

where

r_0 = the resistance at 0°C.
α = the temperature coefficient of resistance, in $\Delta\Omega/°C$.
T = the sample temperature in degrees Celsius.

Even this may not be exact, because the value of α changes somewhat with temperature.

The temperature coefficients of resistance of most solid materials are quite low. They are on the order of 8×10^{-4} to 2×10^{-2} Ω/°C.

The pressure coefficients of resistance (or resistivity) for most solid materials are quite small compared to the temperature coefficients. Thus, we can normally neglect the pressure influence upon the resistivity of a sample of rock. The same idea holds, approximately, for aqueous samples. The pressure coefficient of resistance is substantially larger for compressible materials than for those relatively incompressible and solid. Little work is available on this subject.

Studies have shown a consistent pattern for changes in resistivity with depth. The resistivities of sedimentary surface rocks are generally low to moderate, usually because of their water content. Surface igneous and underlying crustal rocks generally have high resistivities. At greater depths, the conductivities of rocks increase with depth, presumably because of the higher temperatures. This is illustrated in Figure 1.1. In general, specific types of rocks have electrical resistivities which fall into predicable ranges (see Table 1.1). Anomalous values will certainly be found. When they do occur, however, it may be interesting to find out why they occur.

1.1.3 Resistivity Character of Rocks and Fluids

The degree of dissociation in most crystalline structures of rock materials is low. There are few free electrons in an ionic-bond structure; therefore, we would expect a low conductivity (high resistivity) in most rock-forming materials. The exceptions are native metals, sulfides, a few oxides, graphite, and high-grade coals. In other rock materials the resistivity is in the millions of ohmmeters. Tables 1.2 and 1.3 show the ranges of resistivities normal in some rock materials. Rocks and minerals may be divided into classes on the basis of their typical resistivities.

Note that this list does not include the effect of any contained water, pore or bound. For example, quartz is in the very high resistivity range, as are the clays and shales. Yet pure quartz sandstone may have a resistivity range from

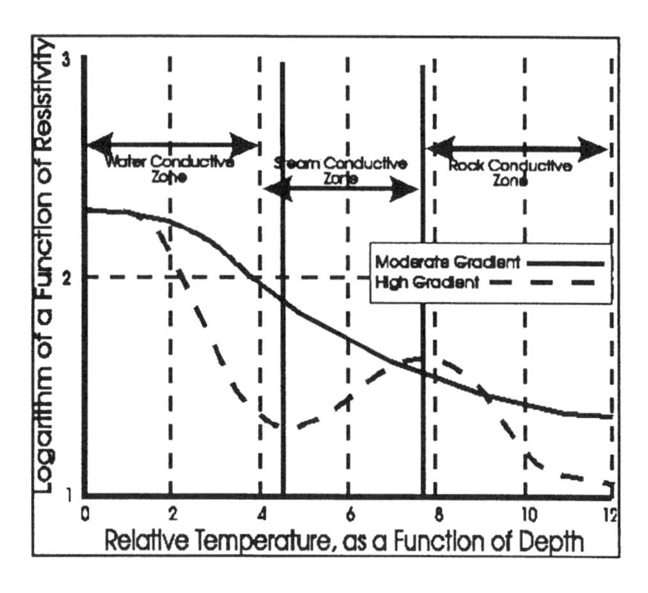

FIGURE 1.1
Approximate resistivity value variations, as a function of depth and temperature curves.

TABLE 1.1
A Table of Several Resistivity Groups

Rock Types	Gradient Range (°C/100 m)	Resistivity Range (Ωm)
Cold, igneous	0-2.5	100→1000
Limestone	0-2.5	45-200
Sandstone	0-2.5	45-200
Cold porous aluvium	0-2.5	6-45
Cold clayey, shaley zones	0-2.5	1-6
Cold, very saline	0-2.5	0.5-1
Hot, dry rocks	2.5-60	30→1000
Moderate temperature zones	2.5-10	4-40
Moderately saline zones	2.5-9	2.5-4
Shallow, warm water zones	10-110	5-50
High temperature zones	10→120	0.5-5
Gas or dry steam zones	60→120	0.5→1000

2×10^{-2} to more than 2×10^{3} Ωm, depending upon the amount and shape of its intergrannular (pore) spaces, the conductivity (salinity) of the pore (or bound) water, and the degree of saturation with gas and/or oil. Clays typically have resistivities around 0.1 to 10 Ωm.

In sediments, the formation resistivity is primarily the function of the fluids in the pore spaces. This is because the conductance of the rock material (quartz, etc.) is in parallel with the conductance of the pore fluid, and the conductance of the rock material is inconsequential. The fluid is the determining factor compared to that of the solid clay mineral.

TABLE 1.2

Representative Resistivity Ranges for Various Materials

Resistivity	Range (Ωm)	Material
Extremely low	$R < 10^{-6}$	Native metals
Very low	$10^{-6} < R < 10^{-2}$	Sulfides, bornite, covelite, micolite
Low	$10^{-2} < R < 10^{2}$	Braunite, ilmenite, marcasite
Moderate	$10^{2} < R < 10^{6}$	Hematite, bauxite
High	$10^{6} < R < 10^{10}$	Cinnabar, scheelite
Very high	$10^{10} < R < 10^{14}$	Calcite, quartz, shale minerals, clays
Extremely high	$R > 10^{14}$	Halite, sylvite, mica

TABLE 1.3

Ranges of Resistivities of Pure Materials

Mineral	Range (Ωm)	Mineral	Range (Ωm)
Anhydrite	10^{7}-10^{10}	Pyrite	10^{-4}-10^{-1}
Galena	10^{-5}-10^{-3}	Pyrolusite	1-10
Hematite	10^{4}-10^{6}	Pyrrhotite	10^{-5}-10^{-4}
Specular hematite	10^{-2}-10^{-1}	Clay minerals	10^{11}-10^{12}
Graphite	10^{-6}-10^{-4}	Sulfur	10^{12}-10^{15}
Calcite	10^{7}-10^{12}	Siderite	10-10^{3}
Sylvite	10^{14}-10^{15}	Biotite	10^{14}-10^{15}
Quartz	10^{12}-10^{14}	Sphalerite	10^{5}-10^{7}
Limonite	10^{6}-10^{8}	Anthracite coal	10^{-4}-10^{-2}
Magnetite	10^{-4}-10^{-2}	Bituminous coal	10^{2}-10^{6}
Marcasite	10^{-2}-10^{+2}	Chalcopyrite	10^{-3}-10^{-1}
Muscovite	10^{11}-10^{12}		

In massive hard rock formations (limestone, basalt, granite, etc.) there is little pore space, and what pore space there is, is likely to be isolated. That is, they probably are not interconnected. Thus, the rock resistance is in series with the pore fluid resistance. In this case, the fluid resistance is negligible, compared to the rock resistance. The rock resistance is the determining factor.

The statistical probable error when measuring the properties of a 3 × 12 in. (7.5 × 30 cm) core is about 1100 times larger than that of the values of a three-electrode focused resistivity system. This problem should always be taken into account.

None of the resistivity systems available from a petroleum contractor are suitable for use in hard rock environments. This is because they are designed for use in petroliferous environments (i.e., low resistivity and resistance). Most petroleum resistivity systems cannot be used above about 1000 Ωms. Mineral systems, on the other hand, are designed for higher resistivity formations (some can be used linearly to 10,000 Ωms or higher). Induction equipment usually becomes nonlinear above 200 or 300 Ωm.

Fault zones and the presence of microfractures in granites (and other hard rock zones) occur on occasion. When this happens, they can frequently be diagnostically detected with a mineral-type focusing electrode resistivity

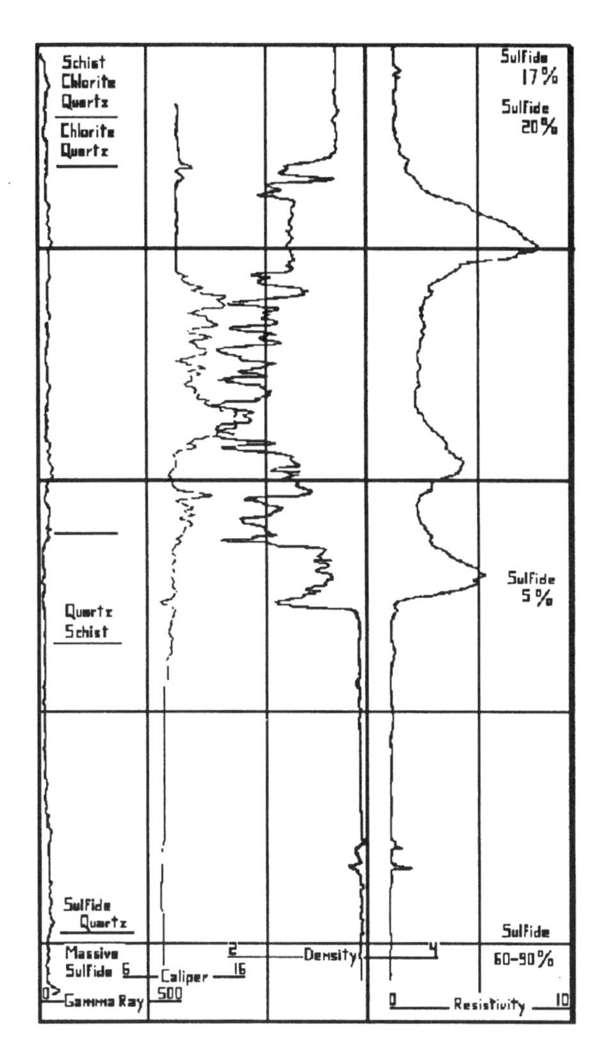

FIGURE 1.2
A Colorado log in hard rock, showing fracturing and massive sulfides.

system. Figure 1.2 shows a log in a Colorado well where the cores showed a hard rock environment, but with extensive fracturing. Note the low resistivity reading, in spite of the high resistivity rock formation.

Similarly, the presence of massive sulfides, as opposed to disseminated sulfides, can be seen with this system. Metallic minerals (metallic sulfides and virgin metals) at high concentrations in hard rock zones can often be detected by resistivity equipment. At low concentrations, the conductive mineral may be isolated by the rock material and the resistivity reading will be high. At higher concentrations, however, the mineral particles may be interconnected, resulting in a very low resistivity. This author found that a concentration above 5%, in one well, could be detected in this manner.

1.1.3.1 Salinity Effects

In addition to the temperature effects upon the resistivity value, there are other things which enter into the picture. Salinity is a common factor. In general, salinity, and thus resistivity, will follow familiar patterns. A given rock type can be expected to have a normal salinity and resistivity range. A rough grouping of such materials is shown in Table 1.4.

TABLE 1.4

Ranges of Resistivity for Various Geological Materials

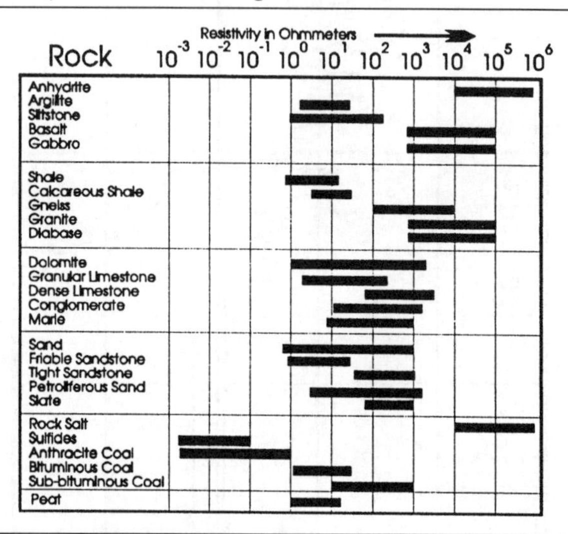

The resistivity of the water in the pore space is given by

$$R_W = \frac{1}{\Sigma(m_a C_a + m_c C_c)} \tag{1.2}$$

where

m = the ionic concentration.
C = the ionic conductivity.

Subscripts *a* and *c* designate anionic and cationic, respectively.
At low concentrations of any specific salt (i.e., NaCl),

$$R_W = \frac{A}{m} \tag{1.3}$$

where

$$A = \frac{1}{\left(\frac{m_a}{m}C_a + \frac{m_c}{m}C_c\right)} \tag{1.3a}$$

The value of A depends upon the specific salt involved. It should be remembered here that most of the "salinity" charts which have been published in the petroleum business have been calculated for dilute solutions of NaCl. Migrating meteoric waters are usually low salinity (moderate to high resistivity). Connate waters are usually fairly to very saline. Marine waters tend to be more saline than continental waters. Note, however, that there are frequent exceptions to this. At extremely high salt concentrations (near saturated solutions), the relationship between salt concentration and resistivity becomes quite nonlinear. This is because the mean free path of the ions has become too short for unimpeded movement.

1.1.4 Checks

As we examine resistivity data (or any data, for that matter) we can make several checks on the validity of the values. Normally, we would expect these values to be "typical" for the circumstances (see Tables 1.1 to 1.5). Of course, we do not know the exact values or there would be no point in making the measurement. If they do not fall within the typical or expected range, the fact may be important and the reason should be determined. It is vital to decide if the measurement is in error or if the values are real before leaving the measurement site. Even if it is impossible or impractical to remake the measurement, the very fact that we know whether the measurement is valid or in error is valuable.

TABLE 1.5

Resistivity Ranges for Some Geothermal Environments

Rock Type	Salinity Range (ppm)	Resistivity Range (Wm)
Clayey zones	100-1000	1-10
Sedimentary zones	100-7000	10-100
Acid Igneous rocks	100→900	100->1000
Alluvial basin deposits	1000-7000	1-10
Geothermal zones	7000-200,000	1-20
Non-geothermal zones	7000-200,000	1-≫1000

There are many things that can affect our measurement and cause aberrant values. For example, we would normally expect a zone of massive granodiorite to have a resistivity above 10^8 Ωm in ambient temperatures below 300°F (150°C). Actually this would probably be beyond the range of the usual resistivity equipment. If the resistivity is lower than that, as suggested by normal readings elsewhere in the zone or borehole, the possibility that the rock is fractured or contains metallic material (i.e., pyrite) should be entertained. Either of these variants can be very significant in a mining or other engineering project, not only as a potential ore body, but, in the case of fractures, because of the potential danger during excavation.

An anomalously low resistivity reading in a massive rock should be investigated, especially if mining is planned. There are several methods available. The microresistivity tools are quite sensitive to the presence of fractures. The microscanner resistivity systems can closely identify electrically conductive fractures in massive rock. The conductive fill material may be water or one of the electrically conductive materials, such as shale, clay, sulfides, or native metal.

If the microscanner indicates conductive fractures, the next step is to identify the fill material. The natural gamma ray system will locate radioactive fill material. This, however, may be clay, sulfides, or deposits of uranium/thorium from water moving through the fractures. A neutron log can identify hydrogenous material (water and/or liquid hydrocarbon) in clays or open fractures. A density system can identify metallic fill.

In the event that metallic material or microfracturing is suspected, the surface and airborne measurements should be inspected. Metallic deposits will probably show up on the airborne magnetic survey. They may even show on the airborne or surface gamma ray maps. Surface resistivity/IP measurements may show either metallic bodies or fracturing (if the latter is extensive).

1.2 Formation Resistivity Measurements from a Cased Hole

Nearly all boreholes, especially in sediments, which are intended for extended use, must be cased to prevent the formation material from caving into the borehole. Some very unconsolidated formations may require casing immediately upon the completion of or during drilling. The presence of the casing material and (to a lesser extent) the annular material (i.e., cement and mud) can also be an error-inducing factor in geophysical measurements. This is especially true of electrical resistivity and conductivity determinations.

Determination of the formation electrical resistivity or conductivity is, of course, a standard, important geophysical method, not only within the petroleum industry, but also for hydrology, metallic mining, structural engineering, and many other disciplines. If it is possible to make the measurements, the determination of the formation resistivity around and adjacent to the borehole is especially interesting. This is because the invaded zone around the open borehole begins to dissipate immediately upon the setting of the casing. Eventually the formerly invaded zone will return to nearly its undisturbed state. Thus, a reliable measurement of the formation resistivity, after the hole has been cased, can have many advantages.

Radioactivity measurements are only slightly impeded by the presence of plastic or metallic casing and annular material. Usually, corrections are easily made. Mechanical wave (acoustic) measurements are also feasible from a cased hole, with a slightly larger difficulty. The most severe restrictions arise with the determination of electrical resistivity and conductivity.

Induction logs use alternating magnetic fields. Thus, they can successfully be run in boreholes cased with plastic or fiberglass. This is because the resistivity of this type of casing is extremely high ($>>10^9$ Ωm) and the induced currents in the casing are minuscule. This type of casing is quite "transparent" to magnetic fields. On the other hand, induction logs cannot be run in conductive casing. Steel casing is very conductive ($>0.6 \times 10^6$ S/m). Aluminum casing is even more conductive, therefore, virtually all of the signal (>99.9999%) in a cased hole will originate from the casing.

Much the same problem is present with electrical resistivity equipment. The metal casing is so conductive that virtually all of the survey current flows within the casing and very tiny amounts flow through the surrounding formation material.

There have been constructive efforts to measure the formation resistivity from within a metal cased hole. A number of methods have been proposed and patented over the last seventy years. Currently, one such device has been field tested and appears promising. It is the Through Casing Resistivity Tool™ (TCRT®) offered by ParaMagnetic Logging, Inc. (Vail et al., 1993).

1.3 Formation Microscanner Tool (FMT)

In *Standard Methods of Geophysical Formation Evaluation*, the standard high resolution resistivity devices are described. They include the microresistivity devices. For the most part, the microresistivity devices are a pad-type, with short, sometimes focused spacings which are isolated from the effects of the borehole by an insulating pad. The electrode array is applied against the wall of the hole. The measurements are usually presented as line graphs of resistivity values as a function of borehole depth. These devices examine very small volumes of the formation, starting at the borehole. A device, employing the same principles, but using a much different philosophy and presentation, is the formation microscanner tool (FMT).

The FMT uses an array of 27 button electrodes, each 0.2 in. (0.51 cm) diameter, on four independent pads, each spaced at 90 degrees from each other. The electrodes are arranged in four rows of seven electrodes and one row of six electrodes. Each row is offset from the adjacent row by 1/5 of a spacing interval. Each electrode overlaps the electrode above it by 50%. The total array is 2.8 in. (7.1 cm) wide by 1.2 in. (3.0 cm) high. Figure 1.3 is a drawing of the pad and array. The other two electrodes shown in the drawing are dipmeter electrodes. The functions of the dipmeters will not be discussed in this volume. The two-pad FMT tool will safely log a 4.25 in. (10.8 cm) diameter borehole. The normal format of the FMT log is a S-W-N-E-S columnar display with depth as the vertical axis.

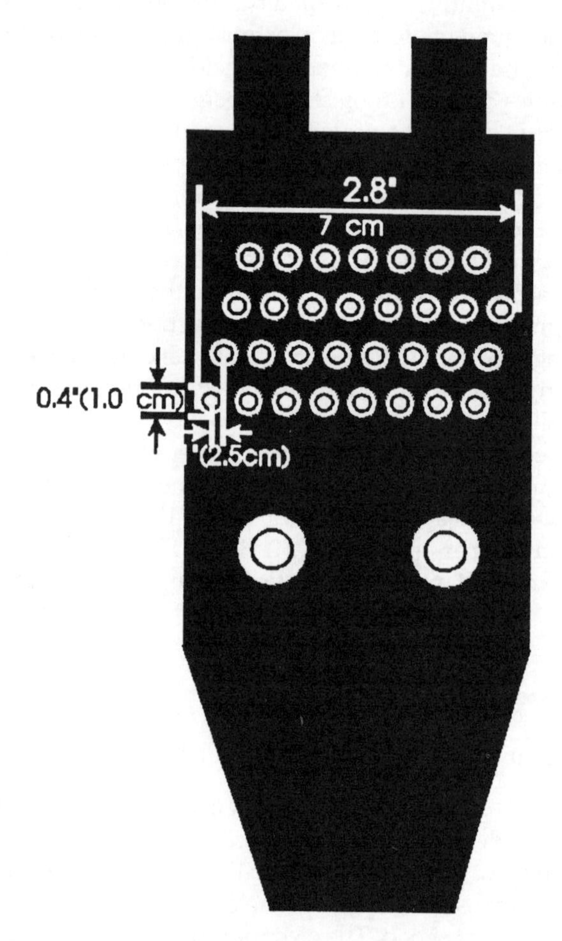

FIGURE 1.3
The face of the FMT pad. (Courtesy of Schlumberger Well Services, Inc.).

The presentation is a gray-scale plot where the degree of grayness (white to full black) is a function of the resistivity. That is, white is low resistivity and black is high resistivity. The position of an information bit (a pixel) is a function of the position of that measurement on the borehole wall. The result of this is a resistivity picture of the wall of the borehole, 3 in. (7.6 cm) wide and the vertical axis is a function of the borehole depth. A copy of this kind of log is shown in Figure 1.4. Presentation can be made with a core photograph for comparison as shown in Figure 1.5. Other logs and/or a core gamma ray log can be used.

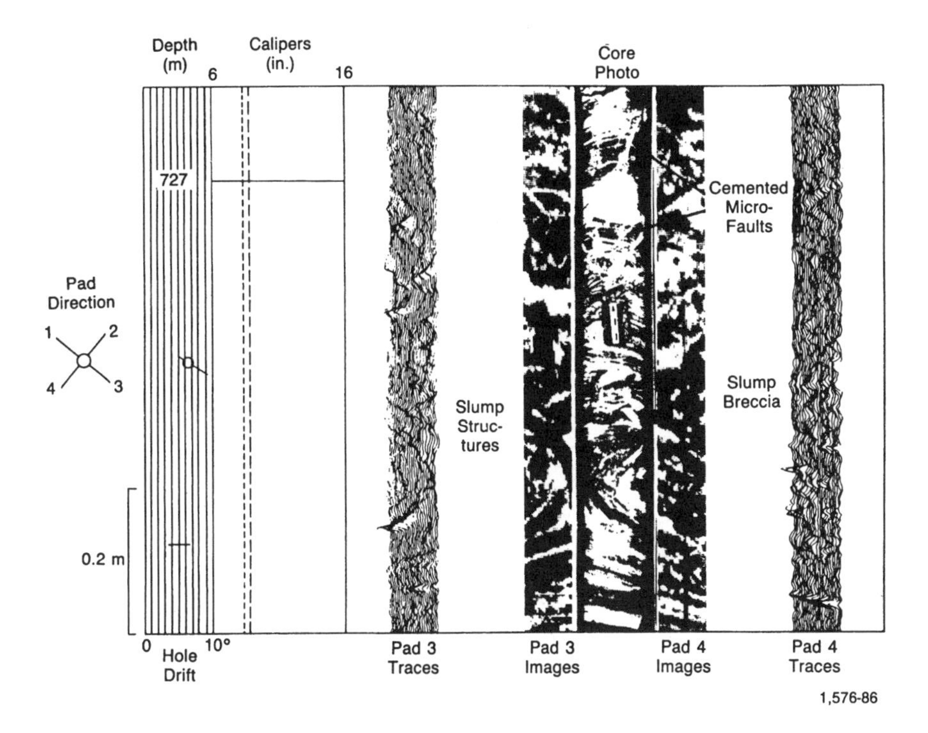

FIGURE 1.4
The FMT presentation. (Courtesy of Schlumberger Well Services, Inc.).

1.3.1 Uses

The FMT is used to detect and examine resistivity features which occur at or near the borehole wall. Some of the uses are listed in Table 1.6.

1.4 The Fullbore Formation Microimager (FMI)

The FMI is a device which is similar to the Formation Microscanner (FMS). The FMI observes more than twice the area, compared to the FMS, and it does this with greater accuracy. It is a large downhole tool and is restricted to boreholes larger than 6.25 in. (15.9 cm).

Both the FMS and the FMI are imaging systems. They produce visual image logs which are pictorial representations of the measured conditions. They can be used qualitatively or semiquantitatively. Discussion of the FMS will also apply to the FMI.

Neither the FMS nor the FMI are effective in fractured or vugular environments where the features have been healed with high resistivity material.

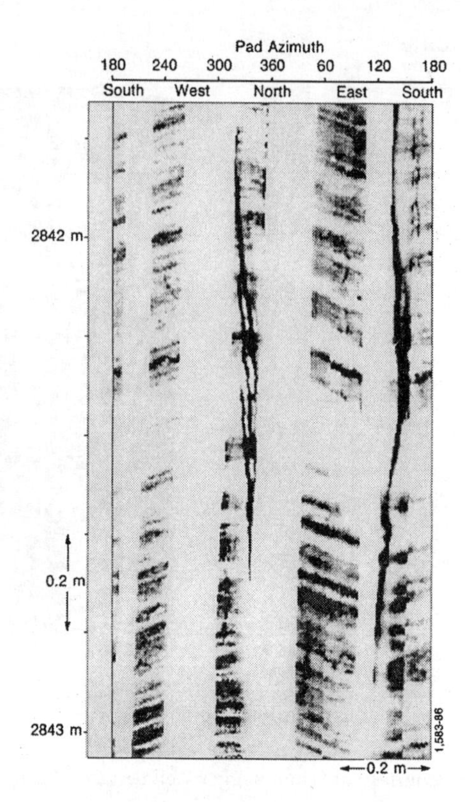

FIGURE 1.5
A full-bore presentation of the FMS log. (Courtesy of Schlumberger Well Services, Inc.).

There must be a resistivity contrast between the host rock and the fill material.

1.5 The Array Induction Tool (AIT)

The AIT contains eight mutually balanced induction arrays. Their spacings range from a few inches to several feet. A single transmitter operates at three different, simultaneous frequencies. Adjacent frequency pairs are utilized by six of the arrays. In-phase (R) and quadrature (X) signal components are measured for each array and for each frequency. This results in 28 induction-array measurements at each 3-in. (7.5 cm) depth interval. The AIT is operated with Schlumberger's MAXIS 500 well-site digital logging unit.

Primary calibrations are made with a set of test loops. Secondary calibrations are made continuously by the central computer while logging. Borehole and depth corrections and depth matching are made and incorporated in the

TABLE 1.6

Uses of the FMT

Flow regime deposition
1. Trough cross beds
2. Tabular cross beds
3. Ripples
4. Very thin single and multiple bed
5. Graded beds
Post depositional bed distortion
1. Faulting
2. Folding
3. Slumping
Non-bed features
1. Fractures
a. Open
b. Healed
2. Vugs
3. Pebbles
4. Concretions
5. Fissures
6. Stryolites

Courtesy of Schlumberger Well Services, Inc.

measurements to produce a set of logs. These logs have median depths of investigation (horizontal) of 10, 20, 30, 60, and 90 in. (25, 50, 75, 150, and 225 cm) from the center of the borehole.

The deep investigation readings are virtually insensitive to the borehole and near-borehole region because of the controlled geometry of the focused measured volume. Similarly, the shallow investigation readings are insensitive to the deep formation region. The median depth of investigation of *all* curves is constant. The median depth of older induction and resistivity systems of investigation depend upon the formation resistivity.

If vertical resolution is defined as the thickness of a bed which will contribute 90% of the total signal at a central measure point, the primary log of the AIT has a vertical resolution of 1 ft (30.5 cm). Two alternative log sets are available: a 2-ft (61.0 centimeter) set and a 4-ft (122 cm) set. The 2-ft set emulates the best resolution of a very enhanced Phasor(c)* log. The 4-ft set emulates the older induction logs. The longer spacings are somewhat less sensitive to sharp changes in borehole size and/or shallow invasion than are the shorter ones.

The logs are matched vertically. They are then deconvolved radially to produce a detailed radial description of the formation conductivity.

---

* Registered Trademark of Schlumberger Well Services, Inc., Houston, Texas.

1.6 Laboratory Methods

Resistivity measurements are done routinely in the laboratory. In the oil and mineral businesses, these measurements are usually made on core samples. In engineering, geology, environmental studies, and real estate, the samples may be cores or they may be other shapes and sizes and from other sources than and including boreholes. Most core laboratories and many university geology laboratories are equipped to perform resistivity measurements.

Laboratory sample resistivity measurements have some good uses which should be carefully considered. Their usual use is for verification of questionable logs. They have some drawbacks, too, and particular problems. One important use, and one which is frequently neglected, is as a teaching tool. An experiment in a freshman physics laboratory can show the concepts of resistance and resistivity as no lecture can. These measurements can also serve as a good research tool.

Laboratory resistivity measurements have tremendous advantage over field measurements in that one has complete control over all of the variables, especially sample dimensions. Cross section, length, volume, density, and temperature are usually controllable or can be measured to any degree of accuracy desired. This is seldom true downhole as well as on the surface in the field. Figure 1.6 diagrams a laboratory resistivity measuring setup.

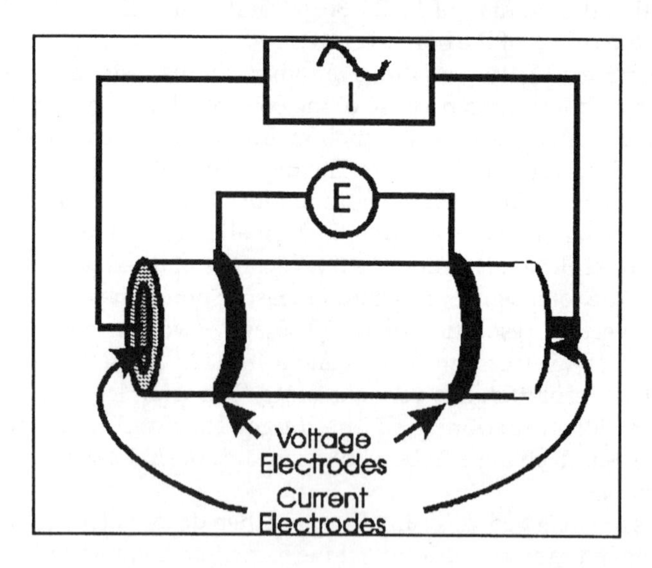

FIGURE 1.6
The laboratory setup for measuring core resistivities.

It is usually feasible to remove the contained (native) water from the sample and replace it with one of known composition and salinity. This is not always done properly, however. The native water must be removed without leaving a residue and without altering the physical and chemical composition of the solids in the sample. The included shales are especially susceptible to this type of alteration. It is also important that the geometry of the dry, contained salts not be changed by the exchange of the pore water. If the various precautions are observed, it becomes possible to determine a number of important parameters.

1.6.1 Samples

Samples can come from drilling cuttings, cores, outcrops, field rocks, cut rocks, or archived samples. Care must be taken to ensure that the sample is competent mechanically and free of any *unusual* inclusions, pores, and fractures. Drilling cuttings, in particular, will be highly contaminated with drilling mud and their clays may be missing.

The native fluid in any of the samples may be dried or partially dried. If possible, a sample of the native fluid should be collected with the sample. This last, of course, is almost impossible with outcrop and archive samples. It may be possible and feasible to transport the samples and to store them in their native fluid in plastic bags. This fluid should be analyzed for composition. If the sample has dried, some of the wash water will have enough of the salts to identify them. The native fluid of cores has probably been partially or wholly displaced by the drilling fluid filtrate. In any event, the fluid in a wet sample will almost certainly be a filtrate (that is, without solids).

Osmosis is certainly a possible error producer. If the laboratory fluid is more saline than the shale, it will tend to dehydrate the shale. If it is appreciably less saline, it will hydrate it. The foreign fluid must be as near the salinity of the native fluid or the shale as possible. The measurement should also be made as soon after impregnation as possible.

Another possible source of error is from the partial solution of soluble crystalline material within the sample. This may be a salt, such as sodium or potassium chloride, or any number of common evaporites. This, of course, will result in a change of the pore geometry, with a consequent error in both porosity and permeability determinations. Again, the solution is to make the impregnating fluid as similar to the native fluid as possible.

1.6.2 Porosity

Interconnected porosity is a major factor determining the resistivity of sediments. It is one of the reasons for running resistivity measurements. Of course, there are many ways of measuring interconnected porosity in the laboratory: gas expansion, density measurements, acoustic travel time, etc. Resistivity measurements are simple and can be quite precise.

The sample must be cleaned and dried without altering the composition and geometries of the solids within the sample. Then it must be completely saturated with a water solution whose composition and salinity are known. The salinity, if the sample contains any clay or shale, must be as close to that of the native water as possible to prevent osmosis from altering the dispersed shales. Vacuum and pressure impregnation will help overcome the surface tension of the water, as surface tension forces can be quite strong. The sample should be dried with warm, dry air. If possible, a lowered pressure atmosphere helps. The saturating water should be as near the salinity of the native water as possible.

Electrode material is not greatly important, as alternating current should be used. Contact materials, such as electrode gel or saturated paper or cloth should be used. Two-, three-, or four-electrode systems may be used. There is an advantage to using an electrode material that has the largest surface area possible. This will reduce the current density at the surface and minimize the electrode surface effects. It is a good idea to stay away from aluminum and stainless steel, as these form highly resistive and tough coatings that can greatly alter the electrode surface character. Iron makes a fair material, as it rusts and forms a large surface area. Lead is satisfactory as its surface coatings are usually nonrectifying. Copper, brass, and bronze can form rectifying coatings that can cause errors. Metal-metal chloride electrodes are excellent, but difficult to use and often expensive.

Either two-, three-, or four-electrode measurements can be made. The two extreme electrodes should be used for the current. The frequency should be kept low, 60 Hz is a convenient frequency. In special cases or for field instrument design, a lower frequency should be tried. Many surface R/IP units use frequencies below 10 Hz. Frequencies above 1000 Hz are used in some commercial instruments, but there is danger of errors, due reactive effects, especially in some shales.

1.6.2.1 Interconnecting Porosity

The interconnecting porosity or effective porosity (ϕ_e) can be used to check on the validity of downhole measurements, especially the choice of the cementation exponent, "m". It can be used to determine the amount of shale in a sand and the presence of isolated porosity. Once ϕ_e is known, permeability, volume, and surface tension and capillary pressure determinations can be made with more confidence. Figure 1.7 shows a hypothetical interconnected porosity environment, as in a uniform sandstone. Low resistivity of the pore water, compared to that of the rock matrix, forms a continuous, but tortuous path, which is what the resistivity system is mainly measuring.

Porosity and permeability determinations, or even estimates, furnish valuable clues for subsurface fluid flow. This becomes important in water wells, particularly in high volume commercial wells, *in situ* leach mining, dam and wall footings, pond and lake construction, and waste disposal projects. Resistivity measurements, backed up by laboratory verifications, make an economical way to handle some of these problems. Both surface and downhole measurements are valuable.

FIGURE 1.7
A diagram illustrating interconnected porosity.

1.6.2.2 Isolated Porosity

The isolated porosity (ϕ_i) is that which is isolated within the rock and has no communication with ϕ_e and is ignored by electrical methods. Because the pores are completely isolated by the high resistivity rock, the resistivity system "sees" essentially only the rock. This value can be determined by measuring the porosity by density and/or acoustic methods. Both of these methods measure total porosity. Thus, if ϕ_e is subtracted from the value of the total porosity, the difference is ϕ_i. This value becomes important in the use of stone as a building material, and it is valuable whenever rock is a load-bearing material. Figure 1.8 illustrates isolated porosity, as in a massive carbonate or a granite.

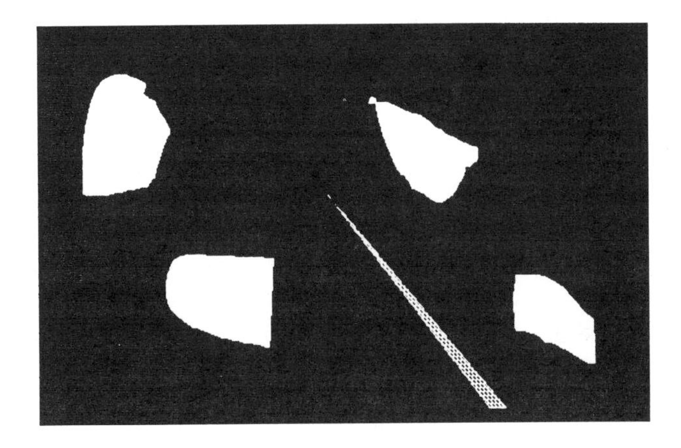

FIGURE 1.8
A diagram illustrating isolated porosity.

1.6.3 Shale Determinations

Laboratory determinations can also furnish valuable data about resistivity of the shale in a shaly sand or carbonate:

1. A gas porosity measurement should be made on the dry sample. The drying should not dehydrate the shale, however.
2. The sample should then be vacuum impregnated with a solution of sodium chloride (NaCl). This solution should have a resistivity or salinity as near to that of the shale as possible to estimate. This precaution is used to reduce shale changes due to osmosis.
3. The resistivity and temperature of the sample should then be carefully measured.
4. Correct the bulk resistivity and the water resistivity for any temperature changes from the initial conditions.
5. Treat the shale resistivity and the known solution resistivity as parallel resistors resulting in the bulk resistivity. Solve for the shale resistivity.
6. Errors will be caused by the shale dehydration during drying, temperature changes, and osmosis with the new solution.

1.7 Oceanographic Uses

Electrical resistivity and conductivity (induction) methods can be and are used in oceanographic studies. They are used both in the water body and in the seabed boreholes. Surface array methods could also be used, but, as far as this author knows, this has not been done.

1.7.1 Water Body

In the water body of an ocean, lake, or river, the electrical resistivity (or its reciprocal, conductivity) are convenient and sensitive measures of the water salinity, either in terms of the equivalent sodium chloride (eNaCl) content or, if the composition is known, in terms of the relative salt contents and the actual salinity.

It is necessary that an accurate measure be taken of the water temperature at the time of the resistivity measurement. This, however, is not a problem, as a temperature log or measurement is routinely made, simultaneously with the resistivity measurement. In fact, a temperature log, run simultaneously with the resistivity measurement, will provide an independent indication of layering and flows. Temperature and ion-type corrections were covered in

Introduction to Geophysical Formation Evaluation. Also, see *Standard Methods of Geophysical Formation Evaluation*, the chapter entitled "The Electrochemical or Diffusion Potential, Effect of Ion Types" for further discussion of ions in solution. The latter also has further discussion of ionic solutions in the same chapter.

Because the total resistivity change from the bottom to the top of a body of water will be relatively small, the sensitivity of the resistivity or induction system can be increased to show small salinity changes. The small salinity and temperature changes can be located, also, by the use of a differential measure or readout circuit.

If small flows, temperature differences, or small layers are to be detected, one of the focusing electrode devices, such as the Laterolog-3, should be considered. Standard oilfield or mineral logging equipment is often used for oceanographic measurements. Almost any of the resistivity or induction systems can be safely used. Note that the single point resistance system can easily be calibrated to measure resistivity.

In a large body of water, such as a lake or an ocean, the type of array used is relatively unimportant. A simple normal resistivity device is as good as a lateral array or a modern induction log. In a small body of water or a river, it would probably be wise to use a shallow depth investigation device to avoid any errors from reading the bottom, top, or banks of the body. Also, in any case, care should be taken to avoid "seeing" the hull of the boat or any wharf or pier. Note that three- or four-electrode lateral devices, the deep laterologs, and the deep induction logs have very large radii of investigation.

1.7.2 Fluid Testers

Laboratory and field testers ("mud testers" in oilfield parlance) are available in several forms. Most of them use electrode arrays and make resistivity measurements. Those that have the electrode array on a central, removable stem are probably better than if the electrodes are built into the wall of the cup. This is because the former is usually easier to clean. Scientific units are usually scaled in conductivity units. Almost all units use a fixed, calibrated volume of fluid, smaller than the radius of measurement. That is, they are calibrated to operate in a *specific* vessel. Many of the simpler units use a two-electrode array. They apply a fixed voltage or potential across the electrodes and measure the current that passes through the fluid. The current is measured and usually calibrated in conductivity units. Most of the more precise units use a four-electrode array. The two end electrodes are the current electrodes, A and B. The two interior electrodes are the measure electrodes, M and N. The current between A and B is held constant and the potential drop across M and N is measured. The unit can be scaled in resistivity or conductivity units. Some of the more complex units use a two-coil induction array. These systems avoid some electrode problems. This type of system is used more frequently in oceanographic laboratories than in oilfield or mineral work.

Fluid testers may have a built-in temperature measurement device. Usually, this is just a mercury thermometer fixed in place in the solution. Some temperature measurements may be made electronically with a bridge circuit, a thermocouple, or a semiconductor sensor. There seems to be no inherent advantage or disadvantage to any of the devices, except convenience. It would probably be wise to check the calibration of any of these devices against a good mercury thermometer.

The units used in oilfield and mineral borehole work are *ohms meters squared per meter* (usually abbreviated *ohmmeters*). Many surface geophysical and engineering units in the U.S. and Great Britain are scaled in *ohms per cubic foot*. Scientific units are usually calibrated in *ohms per cubic centimeter.*

1.7.3 Laboratory Measurements

In addition to resistivity logs, as a function of water depth, fluid samples can be taken and analysis be made in the laboratory. Downhole fluid sample takers should be considered. These are available in single or multiple sample systems and there are many types available.

Oilfield mud testers can be used for oceanographic purposes. They are convenient and available from oilfield supply houses and the major logging contractors. Their accuracy, however, may not always be great enough for scientific purposes. Conductivity meters are available from chemical and scientific supply houses.

Regardless of the type of fluid resistivity tester used, there are several precautions which must be observed, and some of these precautions will depend upon the degree of accuracy desired. Any meter used for scientific purposes (and more exacting than the demands of oilfield work) should be frequently checked with standard solutions. These solutions can easily be made, using ion-free water and the salt mixture of your choice. Both water and salt must be accurately weighed and the solution must be complete. Accurate temperature readings must be made during the calibration. In addition to frequent checking, it is imperative that the electrode array and the vessel or "cup" be clean. This is especially important if low salinity solutions are being measured. Some oilfield units have a cup which uses a squeegee on the end of the electrode array, so that the cup is wiped when the array is pulled out to empty the cup. This, however, is not good enough for careful laboratory work. In this case, both the electrode array and the cup must be carefully washed with ion-free water and dried and inspected before the next use. Finally, *any* fluid resistivity measurement is meaningless without a simultaneous temperature measurement.

2

Photon and Particle Methods

2.1 Introduction

This chapter will deal with some specific parts of electromagnetic radiation and photon phenomena. These are gamma ray, X-ray, and visible radiation.

From the *Handbook of Chemistry and Physics, 61st Edition:* "According to the quantum theory of radiation, the elementary quantity, or quantum, of radiant energy is the photon. It is regarded as a discrete quantity having a momentum equal to $h\nu/c$, where h is Plank's Constant, ν is the frequency of the radiation, and c is the speed of light in a vacuum. The photon is never at rest, has no electrical charge, and no magnetic momentum, but does have a spin moment. The energy of a photon (the quantum of energy) is equal to $h\nu$. Photons are generated in collisions between nuclei or electrons and in any other process in which an electrically charged particle changes its momentum. Conversely, photons can be absorbed (annihilated) by any charged particle."

Electromagnetic radiation is energy, propagated through space or through material media, in the form of an advancing disturbance in electric and magnetic fields existing in space or in the media. The term radiation, alone, is used commonly for this type of energy, although it actually has a broader meaning. It is also called electromagnetic energy or simply radiation.

Electromagnetic spectrum is the ordered array of the known electromagnetic radiations, extending from the shortest cosmic rays through the gamma rays, ultraviolet (uv) radiation, visible, and infrared radiation, and including all other wavelengths of radio energy.

The division of this continuum of wavelengths (or frequencies) into a number of named subportions is rather arbitrary and, with one or two exceptions, the boundaries of the various subportions are only vaguely defined. Nevertheless, to each of the commonly defined subportions there are corresponding characteristic types of physical systems capable of emitting radiation at those wavelengths. Thus, gamma rays are emitted from the nuclei of atoms as they undergo any of several types of nuclear rearrangements. Visible light is emitted, for the most part, by atoms whose planetary electrons are undergoing transitions to lower energy states. Infrared radiations are associated with characteristic molecular vibrations and rotations. Radio waves, broadly speaking, are emitted by virtue of accelerations of free electrons, as, for example, the moving electrons in an antenna wire.

Throughout this chapter two notations for various isotopes will be used. The element name or the standard symbol for the element (i.e., U for uranium or Sn for tin) will be combined with the atomic weight for that particular isotope. For example, the common notation for uranium is the isotope with the atomic weight of 238 atomic units (AU). This may be identified as uranium-238 or as ^{238}U. Both forms are acceptable and commonly used in technical literature.

2.2 Requirements of Gamma Ray Systems

For additional information concerning gamma radiation methods, please refer to Introduction to Formation Evaluation and to Standard Methods of Geophysical Formation Evaluation.

There are four characteristics of gamma radiation measuring systems of which one must be aware before deciding upon a system or a service. These are the reputation of the contractor, and the resolution, sensitivity, and stability of the systems under consideration.

The resolution of a gamma ray system (in common with all measuring systems) is one of the least familiar characteristics of this system. The resolution is the ability of the *system* to define a group of events, from large groups (in some equipment) to individual events (in others). Resolution is "the act, process, or capability of distinguishing between two separate, but adjacent objects, sources of radiation, or between two nearly equal wavelengths" (*Heritage Dictionary of the English Language*). In the case of gamma radiation detection equipment (again, in common with other measuring instruments) the overall resolution depends upon the poorest resolution of the component parts: detector, circuitry, transmitter, transmission medium, receiver, data handling circuit, and output device. All measuring instruments have some or all of these components in one form or another.

In the case of downhole gamma ray equipment, the system consists of a detector, a pulse amplifier, one or more pulse handling circuits, a transmitter, a connecting cable to the surface, a pulse shaper, one or more data handling circuits, and one or more recording devices. The pulse resolution of the system, in the case of most gamma ray systems, is usually the detector. In analog equipment the resolution is also highly dependent upon the time constants of the transmission components.

Gamma radiation consists of a randomly occurring series of photons. These particles arrive at the detector randomly and each has a discrete energy content. This energy content is evident as the frequency (or wavelength) of the photon. Each photon travels at the speed of light of the medium through which it is traveling. Thus, when detecting gamma radiation, we can determine two things: (1) The number of events per unit time and (2) the average of the particle energies detected during that time interval.

Gross-count detectors, such as photographic film, radiation badges, electroscopes, ionization chambers, and gross count scintillators make no attempt to determine energies. They do count the events of a wide energy spectrum during the time interval. This interval may be as long as days or weeks (film, badges, electroscopes, ion chambers), or it may be as short as seconds (ion chambers, Geiger Müeller (GM) detectors, proportional counters, and scintillation detectors). Each type has an important, common use. Resolution is not an important factor in this type of equipment. Sensitivity, stability, and economics are often the deciding factors. Of course, the end use of the information is of prime importance. Sophistication covers a wide range, but is frequently rather low.

Spectral systems measure both the number of events per unit time and the average energy of the impinging particles during that period. Most of the detectors in this category are scintillation detectors of some type. A few are proportional counters, which is a variation of the GM detector. Sophistication of this class is usually much higher than with the gross count systems. Thus, cost, complexity, size, shape, and logistics tend to limit their use.

The purpose of spectral gamma ray equipment is to determine the number of events per unit time *and* the average photon energy during that period. The resolution of the system is important and will depend upon the accuracy of the time period (fairly easily accomplished with modern equipment) and the range of photon energies detected during that time. The latter can be very large or very small.

Commonly, spectral equipment will use one or more detection channels. A channel will have a discrete range of photon energies to detect. The resolution of the channel is the inverse of the range of detectable energies. That is, the wider the channel, the lower its resolution. The sensitivity is the relative number of events the channel will detect, compared to the possible number occurring. The number of channels will vary from one to many thousands, depending upon the system design. To some extent, the resolution is a function of the number of channels. Figure 2.1 shows a typical mineral-type spectral gamma ray (KUT) log.

Thus, we find a wide variety of gamma ray systems that are designed for a large number of different conditions, purposes, and economics. It is easy to use the wrong system for a special purpose.

2.2.1 Accuracy

The limiting factors on accuracy of gamma ray measuring equipment usually are determined by the skill of the design engineer, the quality of the individual components, the purpose of the instrument, the economics involved, and the skill of the operator. Of these factors, we generally have little or no control, or even knowledge, of the first two, but we do have control over the next

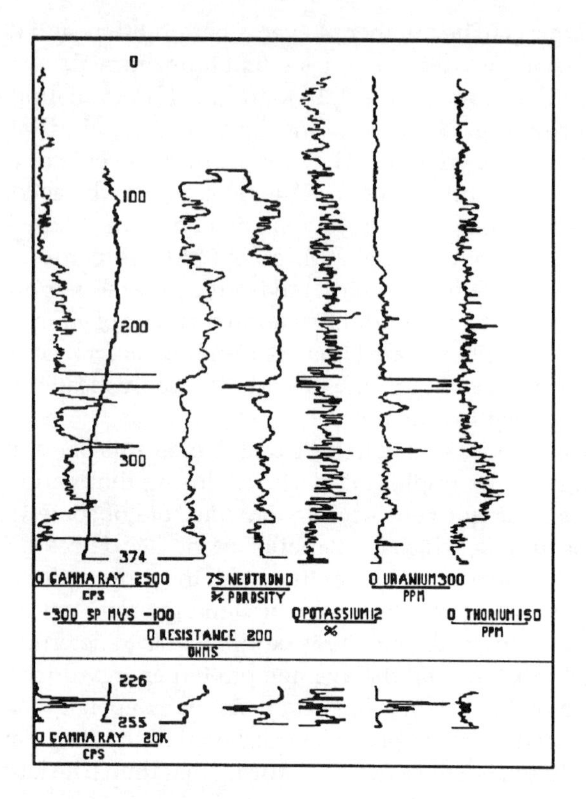

FIGURE 2.1
Extracted from a mineral-type KUT log.

two, and, experience gives us control over the last. This results in an extreme range of end result accuracy. Unfortunately, these factors are seldom considered in evaluating results, especially before the measurement is made. A good operator can get usable results with mediocre or poor equipment. On the other hand, a poor operator usually produces unuseful results from the best equipment. The skill of the operator and the reputation of the contractor frequently take second place to the cost of the operation. This is false economy.

2.2.2 Sensitivity

The sensitivity of an instrument is often the least important of the three criteria, but it is the one most cited. The sensitivity requirement of a system will depend largely upon the amount of time available to make the measurement and its purpose. In other words, a long exposure time can make up for much of the lack of sensitivity. Thus, very sensitive systems are common in airborne and some laboratory work.

Airborne systems are very sensitive because radiation levels are low, flight time is expensive, processing is complex, and interpretation is costly.

Adequate space is usually available for very sensitive detectors and elaborate processing equipment. These systems detect the amount of ^{222}Rn in the air and the radiation from the daughters of ^{235}U, ^{234}Th, and ^{40}K from the ground subsurface. Most airborne surveys are made with gross count systems or three channel (KUT) systems. Multichannel analyzer (MCA) instruments are not used as often because of their lower sensitivity. They are being used more commonly now that solid-state circuitry has reduced the size, weight, and power requirements of the instruments.

Thickness gauges and popular market "GM counters" often do not require much sensitivity because of the high flux rates and the available time for measurement or economic considerations. Most downhole logging equipment usually falls between these limits and is often determined by the available instrument space.

2.2.3 Resolution

The limits of the resolution of an instrument depend, ultimately, upon the resolution of the detector. Of course, high resolution circuitry will not be coupled to a low resolution detector because of the cost of designing higher resolution circuits than necessary. Therefore, it is seldom satisfactory to try to upgrade a system by improving the only detector. The resolution of the thallium-activated sodium iodide detector (NaI) is fairly coarse; however this detector is fairly stable, readily available, and inexpensive. Therefore, it is quite common. In spectral equipment, it has been used with 256, 512, and 1024 channel MCAs. Separation of individual emission lines is often difficult, and usually lies with the detector, not with the MCA.

Other types of detectors have advantages and disadvantages. Denser materials, such as bismuth thalate, offer some improvement. Plastic scintillators do not require sealing, but do not improve resolution. Proportional counters have quite coarse resolution and bulk, and they are generally reserved for special engineering problems.

Ultrapure (or intrinsic) germanium (Ge) detectors have a resolution about 50 times better than NaI. They must, however, be operated at cryogenic temperatures in order to reduce their electron shot noise (that will obscure the pulses). The detector is also very limited in cross-section which limits their sensitivity. It is feasible to use Ge detectors in surface equipment. Ge detectors have been used quite extensively in laboratories. There is little constraint upon size, complexity, weight, and logistics in the laboratory. They have been used very successfully in downhole instruments for the exploration for uranium. The instruments are expensive to build and to operate. Their downhole time is limited by the downhole cooling method which is about 2 to 3 hr in present equipment. Their results in the uranium business have been spectacular. They have resulted in the introduction of several new and improved techniques.

2.3 Natural Gamma Radiation

2.3.1 Laboratory Methods

Most of the laboratory use of natural gamma ray methods is to be found in conjunction with the radioactive mineral industries, uranium, thorium, and evaporite. These methods include quantitative sample analysis, compositional analysis, uranium/thorium ratio determinations, clay/shale determinations, industrial radiography (beyond the scope of this text), core gamma logging, and radon detection.

Uranium (^{238}U) and its daughters decay by a process of photon (γ) and alpha particle (α) emission into several radioactive daughter products, until they finally end in one of the stable or nonradioactive isotopes of lead. The genesis of a new daughter and its decay will eventually result in a stable sequence. The relative amounts of the intermediate daughters will remain constant and only the first (parent) and the last (stable) isotopes will change. This is called "equilibrium". This total process for uranium requires about a million years.

Shale and clay emissions are from gamma, alpha, and beta particles. The gamma and alpha emissions are from the adsorbed uranium and thorium compounds. The beta emissions are from the potassium of some of the clays.

The various emissions from materials containing uranium, thorium, and potassium, and their daughters allow several methods of analysis. These analytical methods are designed to evaluate the present amount of the parent uranium, ^{238}U, ^{235}U, the amounts of ^{226}Ra, and/or the amounts of ^{222}Rn present in a mineral deposit or a geological zone. Many things, in turn, can be determined or deduced from the analytical results of these isotopes.

Basic radioactivity is discussed in *Introduction to Geophysical Formation Evaluation*, Chapter 8. Uranium grade calculations and geological implications of radioactivity using borehole geophysics are discussed in *Standard Methods of Geophysical Formation Evaluation*, Chapters 6 to 10. Our discussion here will be limited to surface and laboratory use and procedures.

2.3.1.1 Core Gamma

An excellent device for correlation for *all* evaluation procedures is the core gamma device. It is simply a moving belt, within a highly shielded enclosure, with a gamma ray system and a color camera. Many of the devices have provisions for entering core information, such as depths, into the record, with the gamma ray log and photographs. The purpose of the device is to provide a close correlation between logged depths and core depths. The gamma ray log

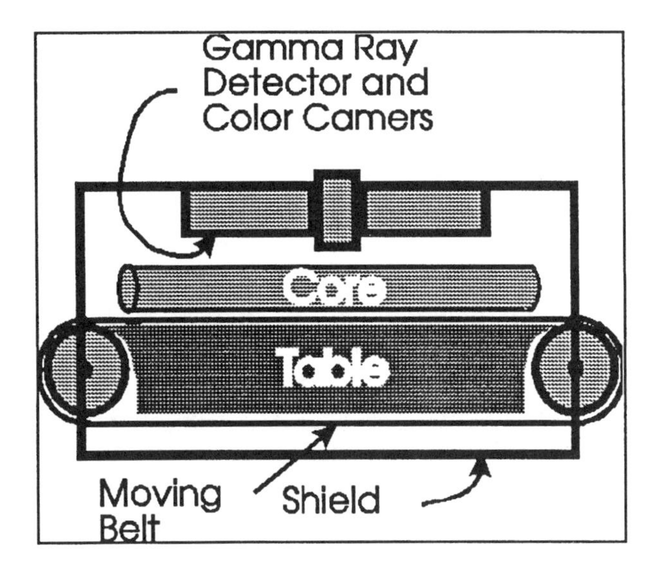

FIGURE 2.2
The core gamma device.

from the core gamma device can be matched closely with a downhole log from the same hole. Figure 2.2 is a diagram of a core gamma device.

The core gamma devices are designed as correlation devices. They should not be used to do quantitative mineral assaying without a specific calibration. The problem is that the geometry of the measurement with the core gamma unit has not been considered because it is not necessary to do so in a qualitative, correlation unit.

If the core gamma unit is to be used to assay a radioactive mineral, the unit must be calibrated for that mineral and use. Two or more samples of the mineral of different grades, the same size, shape, length, and density as the sample, must be used to calibrate the unit. A response curve should be drawn to identify the grade of the unknown.

The important thing is that the geometry of the measurement is of prime importance to a quantitative radioactive measurement. Remember, the core gamma unit probably was designed only for correlation.

2.3.1.2 Beta-Gamma Method

A laboratory method to determine the presence of disequilibrium in a sample suspected to contain uranium is the beta-gamma method. This method reads the relative amounts of high energy beta emission from a sample and the high energy gamma emission. The beta emission is primarily from protactinium, ^{234}Pa; bismuth, ^{214}Bi; and lead, ^{214}Pb. The gamma emission is primarily from bismuth, ^{214}Bi, and lead, ^{214}Pb. Subtracting the amount of gamma emission, after applying a suitable factor, from the beta emission leaves, essentially, the emission from the protactinium, ^{234}Pa. The half-life of ^{234}Pa is about 2.18 min and that of the intervening ^{234}Th is only 24 days. Thus, the ^{234}Pa has a very high probability of being in equilibrium with the original ^{238}U and thus, indicates the presence of ^{238}U.

2.3.1.3 Closed Can Analysis

A closed can analysis is used when disequilibrium, caused by the escape of radon, is suspected. Because the half-life of ^{222}Rn is about 4 days, sealing a sample in a sealed can to allow the radon to accumulate will show the presence of that type of disequilibrium in a uraniferous sample. Figure 2.3 is a diagram showing the construction of the "closed can".

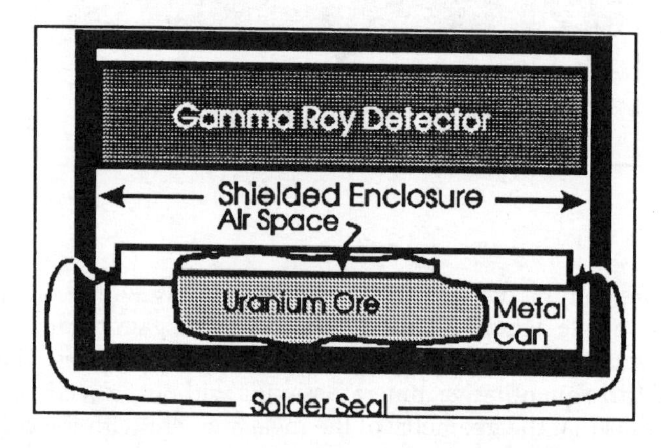

FIGURE 2.3
The closed can gamma method.

2.3.2 Estimation of Uranium Amount

Because uranium (^{235}U) is a valuable fuel mineral, much effort is spent exploring for and evaluating deposits of uranium. At this time there is almost no uranium exploration in the U.S, although it does continue in other parts of the world. Many of the methods are directly applicable to the search for other fuel and nonfuel minerals. For example, potassium evaporite exploration methods and analysis can use many of the processes developed for the search for uranium. Once a deposit has been found and valuable amounts of the mineral have been determined from geophysical surveys, logs, and geological evaluations, the grades of the mineral in numerous places and the average deposit grade and probable mineral amount must be determined, verified, and mapped. Verification is done with core samples and other types of samples in the laboratory. These are then compared with the indications from the logs and extrapolated over the whole deposit. If the deposit is economically minable, it becomes an ore body.

Fissionable ^{235}U occurs in natural mineral in an apparently fixed ratio with ^{238}U. ^{235}U is 0.72% by weight of the ^{238}U in most present-day deposits. Because both ^{238}U and ^{235}U decay radioactively at different rates, this ratio actually is constantly changing. Refer to a chart of the decay series of radioactive materials. The half-lives of the two isotopes are long enough so that this ratio holds commercially, for the present.

Because fission depends upon the mass of the fissionable isotope present in close proximity, one can envision a deposit, in the past, where the amount of ^{235}U was great enough that spontaneous, natural fission could have occurred. This apparently happened in a few (very few) deposits. This is indicated because, in these few deposits, the ratio of $^{235}U/^{238}U$ is much less than the usual 0.72%.

Uranium is a weak gamma emitter. All of the natural isotopes of uranium, ^{238}U, ^{235}U, and ^{234}U are primarily alpha particle (α) emitters. These particles are easily stopped, as they are large. Thus, they do not penetrate far through the formation material. In fact, a heavy grade of kitchen aluminum foil will stop the majority of those that are natural. Therefore, gamma radiation is the most commonly used borehole and surface radioactivity method for detecting the presence of uranium. Because the alpha particle is the nucleus of the helium atom, the presence of helium gas indicates the possible presence of uranium or thorium at present or sometime in the past.

The bulk of the detected gamma rays come, not from uranium, but from its distant daughters, bismuth-214 (^{214}Bi), lead-214 (^{214}Pb), and thallium-208 (^{208}Tl). In addition, the fissionable isotope of uranium, ^{235}U, occurs in such relatively small amounts that it is difficult to detect directly. Therefore, the gamma radiation from a sample of the daughter product, ^{214}Bi, is used to quantitatively measure the amounts present of ^{238}U and ^{235}U. This type of assay is quick, convenient, easy, and inexpensive to perform downhole, on the surface, from the air, and in the laboratory. If it is done carefully and properly, it is quite accurate. There are, however, many problems associated with it that must be addressed. These problems are caused by the relative distance between the uranium and its gamma-emitting daughter and the properties of the uranium compounds and the several steps in between the parent and the end daughter. The average time for ^{238}U to decay through twelve steps to ^{214}Bi is approximately 10^6 years, if nothing happens to disturb the sequence of events during that time. See Figure 2.4 which is a typical mineral-type calibrated gamma ray log.

If the counting rate from the gamma-emitting daughter is directly representative of the amount of uranium present, the sample is said to be in equilibrium. That is, the rate of production of each step is exactly the same as its rate of decay.

One of the daughters of ^{238}U is radon-222 (^{222}Rn). It is the eighth step between ^{238}U and ^{214}Bi. Radon is a gas and the isotope ^{222}Rn has a half-life of approximately 4 days. If the radon escapes the formation during three or four half-lives, then there will be no ^{214}Bi to emit gamma rays, although there may be uranium present. This sample is out of equilibrium.

Many of the compounds of uranium are quite insoluble when they are in a reduced form. If they are oxidized (as by oxygen-bearing ground water or

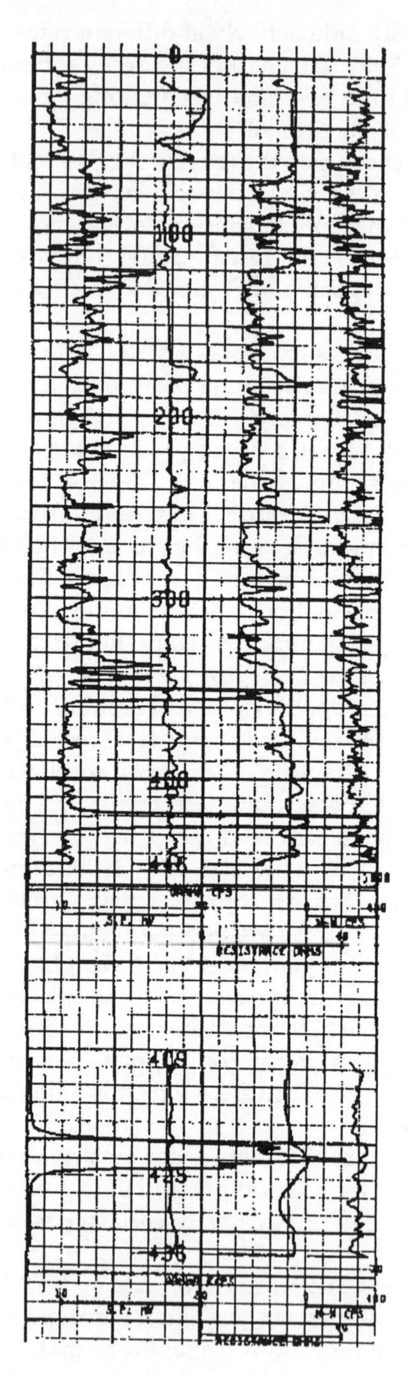

FIGURE 2.4
A mineral-type calibrated gamma ray log.

exposure to the air) they can be very soluble. In some cases, the solubility from reduced to oxidized can change from a few micrograms per liter to grams per liter. Thus, after being in place for a long enough time to have extensive daughter products, ^{238}U can be oxidized, dissolved, and transported away. Thus, there may be good gamma ray indications, but poor uranium presence. This is also out of equilibrium, but in the other direction.

The way around this problem is to perform chemical analyses, X-ray analysis, or other verifying tests on some of the laboratory samples. If these verifying samples are chosen from representative locations, a probable representative equilibrium factor can be established for that area. This factor can then be applied to the log values, core samples, and the surface and (perhaps) the airborne values with confidence.

Thorium compounds can, and often do, occur with uranium deposits. This is especially true with hard-rock deposits and thermal sedimentary deposits. Thorium is also radioactive and has good gamma-emitting daughters. Thorium can be tolerated and mixed with uranium, in some reactors. Canadian reactors typically burn a mixture of uranium and thorium. Other reactors, such as those used in the U.S., must have a much purer grade of uranium. Therefore, the relative amounts of thorium in the deposit can have a bearing on its value.

Thorium, like uranium, is not a good emitter of gamma rays. The daughter product normally used to detect thorium is thallium, ^{208}Tl. Its end product is stable lead, ^{208}Pb. The common radon isotope of thorium is ^{226}Rn, which has a half-life of only 54 sec. It is much less likely to escape than is ^{222}Rn. Also, thorium compounds, in general, are not as soluble when oxidized as uranium compounds. Therefore, disequilibrium is not as great a problem with thorium deposits as it is with uranium deposits.

2.3.3 Small, Portable Radiation Detection Units

Small portable radiation detection units include hand-held field and laboratory units, Geiger counters, and monitors. These small units have become an important and necessary tool for the field geologist, engineer, and for laboratory use. There are several types available and units have been designed for a number of purposes. See Figure 2.5 which shows approximate layouts for two types of small hand-held units.

Hand-held gamma ray and X-ray units will measure the radiation field strength *at the detector.* These units should never be used for quantitative purposes unless precautions and calibrations have been made. They may be used with confidence, after calibration, to measure the radiation field strength at a specific point. The reason for this is that the geometry of the measurement enters into the calibration of the instrument. Thus, hand-held units may be used to map the radiation field strength of an area by measuring at many individual points, at a specific detector orientation and distance from the

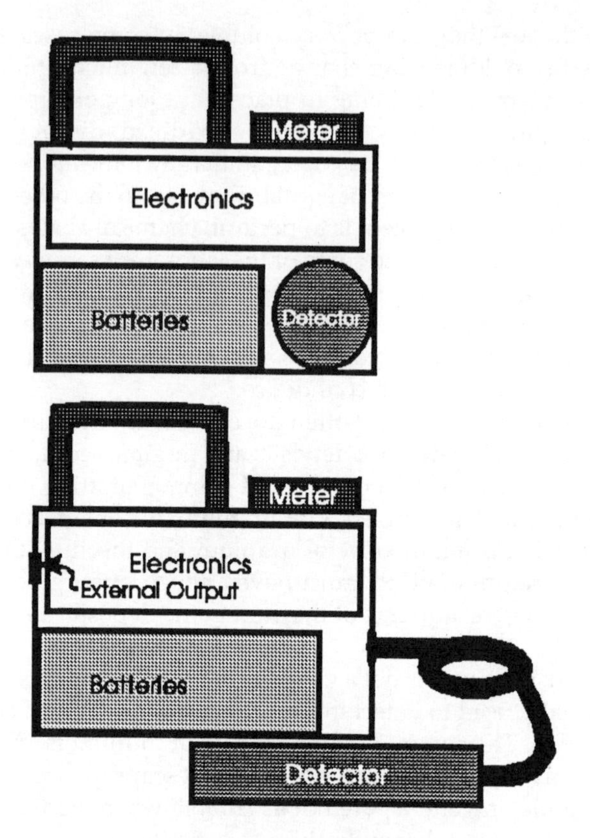

FIGURE 2.5
Diagrams of two types of hand-held gamma ray instruments.

radiator (i.e., each measurement at 4 ft above the ground). Then isorads (lines of equal radiation) may be drawn in the usual fashion.

If the hand-held unit is used to check mineral samples, it must be calibrated in the manner described for the core gamma device in the previous section. It is quite feasible to make comparative measurements of samples by making sure that the position relationship of each sample and the detector remains the same.

Hand-held radiation detectors are valuable for personnel safety monitoring. At least one unit must be available whenever discrete sources are being used. Fields should be checked near personnel areas before, during, and after using the source.

Normal practice, in laboratories where sources are frequently used, is to have a unit mounted in a prominent place in the work area. A relay on the output can be used to light a warning sign and/or sound an audible alarm whenever the radiation level exceeds a preset safe limit. This type is particularly effective if a simultaneous permanent record is made of the field reading and of the times and dates.

An important use of hand-held units is in contamination detection and cleanup. For these purposes, the unit must have sufficient sensitivity and stability to locate trace contaminated areas and objects. It is vital that the unit itself not be allowed to become contaminated. It may be wise to enclose the unit in a light, disposable plastic bag. It is important that the contaminated material be measured and recorded in a separate, calibrated, shielded unit. Such material should be packed and sealed and its contents labeled clearly. It should then be shipped to an authorized disposer of such hazardous materials.

The major field use of hand-held units, of course, is to look for anomalous radiation areas. These areas may indicate the proximity of interesting amounts of radioactive minerals (uranium, thorium, and potassium compounds) and warrant further investigation. The type, ruggedness, and sensitivity of the unit will depend upon the type of investigation.

2.3.4 Detector Types

Detector types have been covered in *Introduction to Geophysical Formation Evaluation*. A few additional comments are interesting, however. Refer to Chapter 10, "Transducers and Electrodes", for additional information on detectors for radioactivity. These gamma ray detectors are also used for X-ray detection.

2.3.4.1 Geiger-Müller Detectors

The original hand-held gamma ray radiation units used Geiger-Müller (GM) detector tubes. These were used by the thousands in the early hunt for uranium mineralization. This was because they were cheap, rugged, and "everyone had one". This is still called the Geiger counter.

GM detectors are rather insensitive and lack photon energy discrimination. All detector pulses are identical regardless of the incident photon energy. The pulse height depends only upon the detector design and the exciting voltage across it. GM detectors have long deadtimes (the length of time, after the initiation of one pulse, before being able to detect a second event; 100 μs or more). This is because of the cascade action of the event within the gas of the tube, making the GM tube unacceptable for some uses. On the other hand, it is simple and rugged.

Sensitivity of the GM detector is increased by operating several tubes in parallel, in a bundle. The individual tubes in the bundle must be isolated electrically from each other by resistance, capacitance networks so that a pulse in one tube will not fire the rest of the tubes. If this network is designed properly, the sensitivity will be increased and the deadtime reduced proportionally to the number of tubes used. Figure 2.6 shows the essential construction of a GM or proportional counter tube.

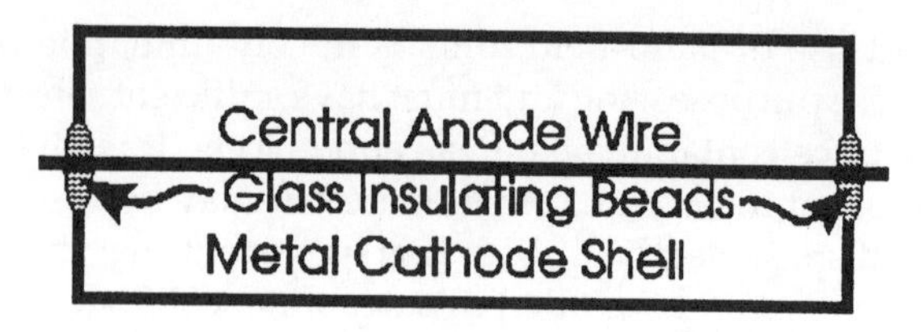

FIGURE 2.6
A diagram of a GM or a proportional detector tube.

2.3.4.2 Proportional Detectors

Proportional counters are similar in construction to the GM tubes, but they are operated at a lower exciting voltage (below the critical GM voltage). The result is that the ionizing events do not cause a cascade, but ionize a path through the gas filling the tube. This sheath of ionized gas migrates to the anode wire in the center of the tube. As it reaches the anode, it changes the anode potential slightly and in proportion to the amount of ionization. This change of potential is a function of the energy of the impinging event.

This type of proportional counter is seldom used today, with one important exception. The ^3He neutron detector is basically a proportional counter. It is, however, usually used in a nonproportional mode.

2.3.4.3 Scintillation Detectors

Modern hand-held detectors almost universally use scintillation detectors to detect gamma and X-radiation. This type of detector can be much more sensitive than GM detectors. They also have much shorter deadtimes ($<10\ \mu s$). Therefore, high radiation fields and increased sensitivity are not as serious problems as with the GM detectors. Their disadvantages are that they are much more delicate than GM tubes, more expensive, and require more circuitry. Modern jacketed, sealed crystal/photomultiplier tube combination units are reasonably rugged, however.

The sensitivity of a scintillation detector depends upon the cross-section area of the crystal, normal to the direction of radiation. It also depends upon the type of material used for the crystal. The type of crystal has some bearing upon the ruggedness of the unit. Most units use the thallium-activated sodium iodide (NaI) crystals as scintillators. Other types of crystals and specialized rugged assemblies are now available. Plastic crystals do not need to be hermetically sealed. Bismuth thallate crystals are denser than sodium iodide and therefore can be better for spectrographic work. Figure 2.7 shows a rough outline of a modern scintillation detector.

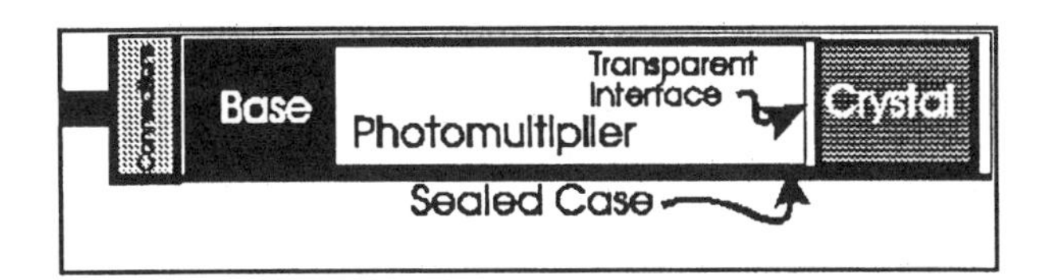

FIGURE 2.7
A diagram of a typical sealed scintillation detector.

2.4 X-Ray Methods

The value of X-ray methods for mineral exploration lies in the fact that X-ray fluorescence consists of emissions whose wavelengths are characteristic of the activated atom. Further, the type, shape, and location of crystalline minerals can be evaluated with X-ray techniques.

X-rays are defined as soft or low energy electromagnetic energy. They are indistinguishable from gamma radiation except for their energy content (wavelength) and their origin. Gamma radiation originates from nuclear processes. X-radiation usually originates from electron interaction.

X-rays are generated by the interaction of energetic electrons with the K, L, and M orbital electrons of atoms. The K, L, and M shell electrons receive energy from the impinging electrons and are displaced from their normal orbits. Later, upon losing that energy, the emitted energy is in the form of X-rays that have wavelengths characteristic of the atom from which they were emitted.

Commercially, X-rays are generated by accelerating electrons with an intense voltage (potential) field to strike a heavy metal target. The usual target material is tungsten, element no. 74. If the electron energy is high enough to allow the electrons to penetrate to the lower orbital shells, characteristic X-rays will be emitted. Figure 2.8 show a diagram of one type of modern X-ray generating tube.

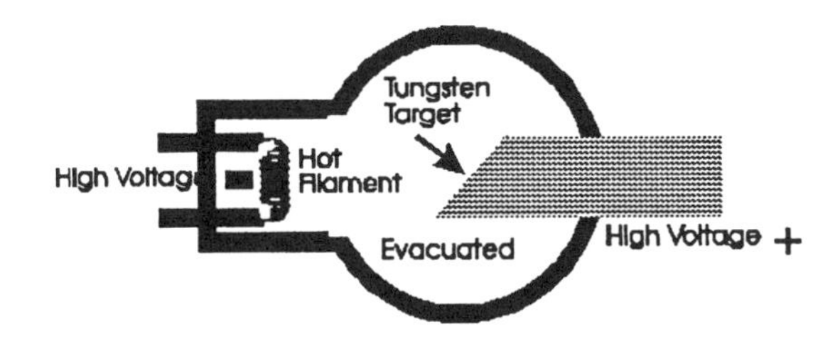

FIGURE 2.8
A diagram of one type of X-ray generator.

2.4.1 Borehole Logging Systems

The use of X-ray logging has been tried successfully, but is not in general commercial use at this time. Field experiments were carried out in Germany and reported at a meeting sponsored by the International Atomic Energy Agency (IAEA) in Vienna, Austria, circa 1980. The purpose was to locate deposits of uranium. The test achieved its purpose, but there were severe restrictions. The system was very sensitive to the rugosity and the wetness of the borehole wall.

The equipment in the German experiments consisted of a low energy gamma ray detector and a source of beta particles (the type of source is unknown). These were run on a conventional mineral-type logging truck. The depth of the borehole is unknown, but it was shallow, uncased, and air-filled. Because the source of activation was beta (electron) emission and the maximum X-ray energy was only slightly above 100 keV, the borehole was necessarily uncased and gas-filled.

2.4.2 X-Ray Diffraction Analysis

X-ray diffraction analysis is commonly used in conjunction with visual identification of samples (i.e., core samples). X-ray diffraction lacks the sensitivity of visual identification, but it is more diagnostic. It does not convey surface textural features, as does visual inspection. Diffraction analysis is a standard technique in the exploration for uranium.

2.4.3 X-Ray Crystallographic Analysis

X-ray crystallographic analysis is a valuable laboratory method for determining the mineral composition of samples. This method makes use of the occurrence of characteristic microcrystals of the various mineral compounds in samples, such as cores and cuttings. These crystals are generally very small and randomly oriented. They are also generally in an opaque or semi-opaque body or matrix. The dominant compound crystals will show a characteristic pattern on a detector due to the total reflection by the facies of crystals lying in positions to expose facies at critical angles.

X-rays are used for this analysis because their short wavelengths and penetrating ability allow them to penetrate material surrounding the microcrystals and reflect from their surfaces. If an electromagnetic wave strikes a crystal face at a critical angle, it will reflect at that characteristic angle. Thus, numerous randomly oriented crystals will generate a characteristic pattern on a surface normal to their mean path. The major pattern will be that of the dominant crystal type within the sample. The information to identify the crystal from its pattern can be found in numerous handbooks (i.e., *The Handbook of Chemistry and Physics, 61st Edition*). Professional analytical laboratories will, of course, have more complete listings. Various types of detectors are used, such as photographic films and fluorescent screens.

2.5 Radon Surveys

The free radon content of the soil varies widely. The presence of radon depends upon a number of factors, including the air temperature, changes of temperature, ambient air pressures, pressure changes, the presence of disseminated radioactive minerals, the amount and distribution of clay minerals, the length of the migration path, and the presence or lack of anomalous uranium deposits. It is not possible to assign a unique radon value to an area. Probable values, comparable values, and trends, however, can be successfully used.

In one series of tests (Bhatnagar, 1973), multiple boreholes were used in each of fourteen locations. Some of the locations were known to have no mineralization. Others were known to have above background uranium mineral deposits. In each borehole, a phosphor-coated chamber was lowered into the air-filled borehole. Borehole air was circulated through the chamber. The chamber was then sealed and brought to the surface. The radon radioactivity of the contents was measured. Log-normal plots were made for each location.

The plots of the barren area boreholes showed a simple, single distribution. An example of one of these is shown as Figure 2.9. Those from mineralized locations tended to show bimodal distributions. Figure 2.10 shows an example of the presence of radon and radioactive mineral deposits.

In a further test, soil samples were taken from around a borehole in a barren area. A single distribution was found for the radon contents of the samples. A radium chloride source was lowered into the borehole. After several weeks, soil samples within 1 ft (30 cm) of the borehole began to show the bimodal distribution. Those farther out remained a simple single trend.

The use of radon measurement for uranium exploration was suggested by Behounek in 1927. He studied the anomalous radioactivity he had found in springs, soils, and in the atmosphere around deposits in Joachimthal, Czechoslovakia. Field prospecting was attempted by the Soviets in the early 1930s, using soil air. Further studies were made when uranium exploration became more important after 1945, but uranium was not widely used prior to 1960.

Because radon is an inert gas (chemically), it can migrate freely through porous rock and soil without reacting with local materials. It occurs as ^{222}Rn from ^{238}U, ^{220}Rn from ^{220}Th, and ^{219}Rn from ^{235}U. It diffuses through the enclosing or parent material after forming by decay. It is, however, unlikely to escape from the mineral grain unless the parent radium is close to the grain surface. This can occur if the grain is small or if the grain has a thin crust of radium from hydrothermal or weathering alteration.

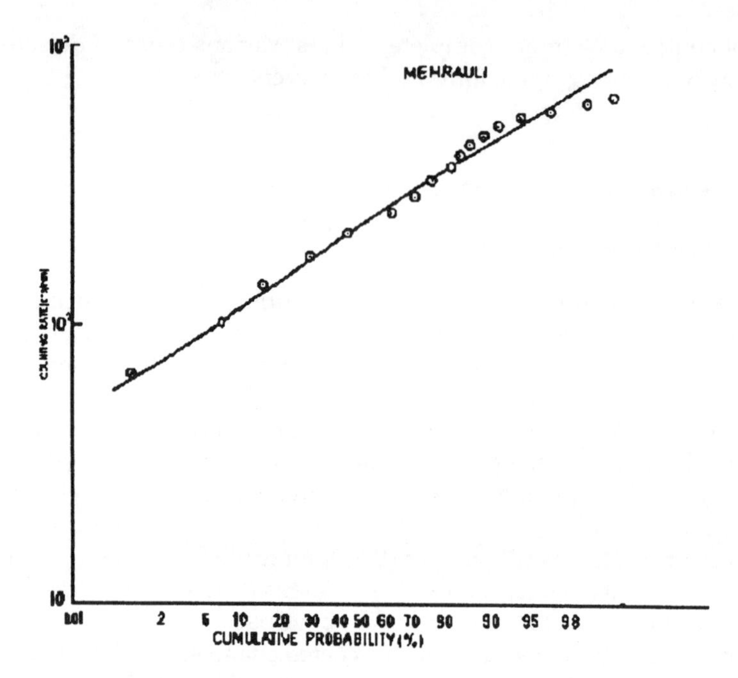

FIGURE 2.9
The probability diagram of the presence of radon over a known barren area.

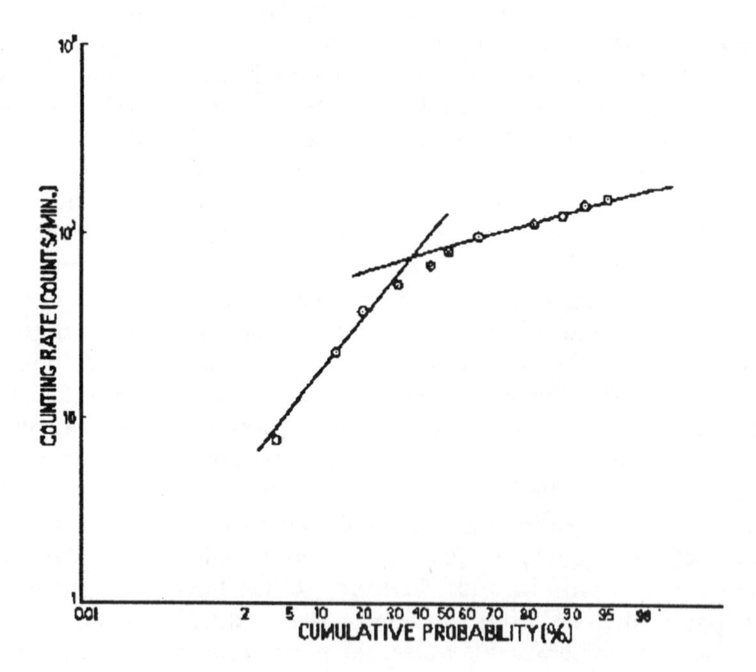

FIGURE 2.10
The bimodal response of the probability diagram over a known mineralized field.

After the radon escapes from the mineral, it will diffuse through the ground air and water and will dissolve in the water to a small extent. With a half-life of nearly four days, ^{222}Rn can diffuse farthest from its source. It is unlikely that ^{220}Rn, with a half-life of 52 sec, nor ^{219}Rn, with a half-life of 3.9 sec, will be found outside the immediate vicinity of their sources.

Arid soils have almost complete continuity between ground and atmospheric air. Thus, low atmospheric pressures and strong winds tend to draw the soil gases upward and reduce the radon concentrations in the soil. Calm, high pressure conditions and wet top soil tend to restrict the flow from the soil and allow radon to concentrate in the soil. Humic top soil can seal off the top thin layer and result in a concentration of radon near the surface.

The diffusion length of ^{222}Rn in water is severely limited. Groundwater flow, even in the neighborhood of springs, is low; on the order of 1 ft (30 cm/day). Studies in Canada (Dyck and Smith, 1968) show that most of the radon found in surface waters is derived from radium absorbed in bottom sediments or from springs feeding lakes and streams. Complete saturation of the soil (with water) limits the diffusion length of ^{222}Rn to about 10 cm (4 in) in water. Therefore, any significant migration of ^{222}Rn will depend upon the flow of the water bearing it. The charts of Figures 2.11 and 2.12 show the effects of atmospheric conditions upon soil air and humic conditions, respectively.

Radon surveys of soil air have their greatest potential in areas where uranium deposits are known or suspected to exist. Note also that, because radon

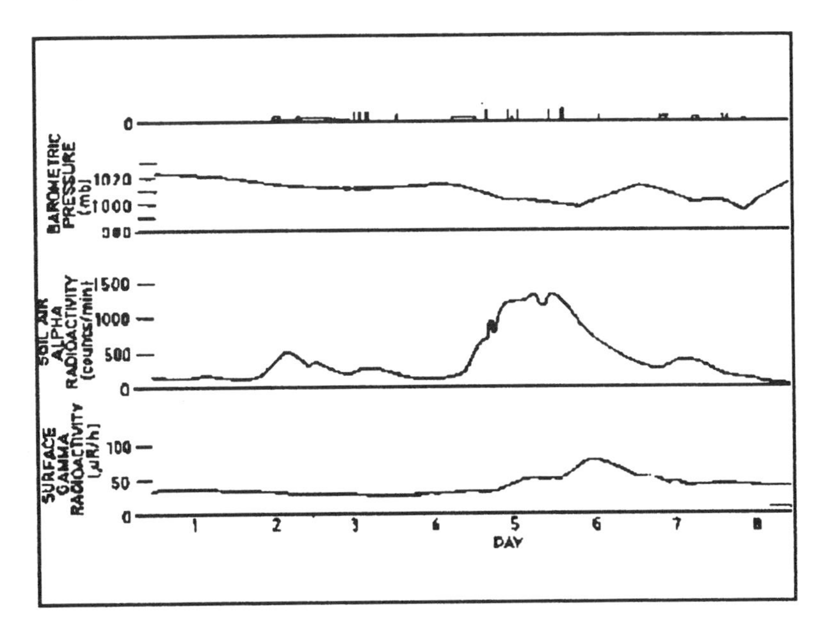

FIGURE 2.11
The effect of arid atmospheric conditions on the distribution of radon over a mineralized zone.

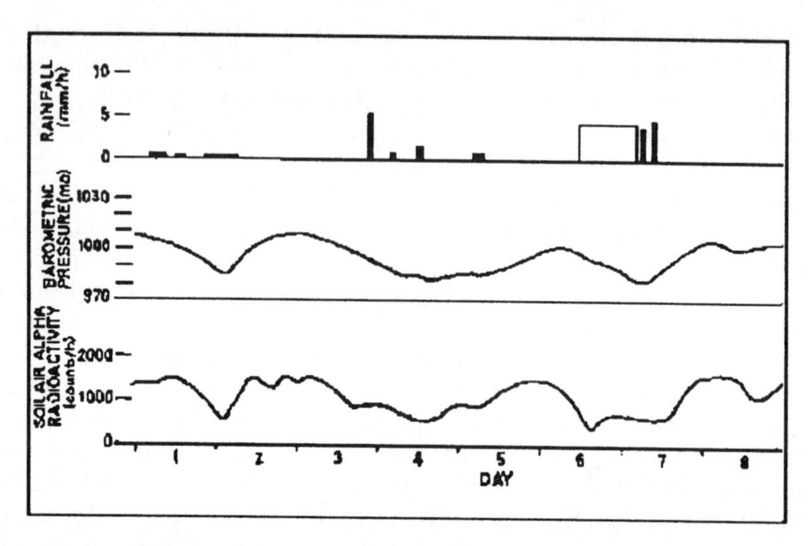

FIGURE 2.12
The effect of rainfall on the presence of radon over a mineralized zone.

and other radioactive elements decay by the emission of alpha particles (essentially helium nuclei), helium will be present in anomalous amounts in the neighborhood of uranium and thorium deposits.

2.5.1 Alpha Cups

During the decade of the 1960s, a radon survey method was successfully used for uranium exploration that proved quite valuable and relatively inexpensive. The method used radon or alpha cups. This method makes use of the radon (^{222}Rn), which is one of the important daughters of the uranium (^{238}U) decay. Because radon is a gas, some part of it may migrate toward the surface above a uranium or thorium deposit. Since ^{222}Rn has a half-life of about four days, there is a substantial probability that some will reach the surface. ^{222}Rn is an alpha particle emitter when it decays to astatine. A photographic film, in the path of an alpha particle, will show a track marking the path of the particle. After the film has been developed, the number of tracks can be counted. This, and more sophisticated techniques, are common in the radiation protection badge industry. In the alpha cup method, the film is enclosed in a thin, light-tight cover and fastened in the bottom of a cup. Hundreds or thousands of these cups are placed in shallow depressions, inverted in a grid, and covered with earth and left for several days. After development and counting, the values are plotted on a map and the isorad lines drawn. The isorads will tend to encircle any radioactive deposit. The intensity will be a function of the mineral amount, type, and overburden type. This method is generally qualitative only.

The alpha cup method depends upon some of the ^{222}Rn being able to escape the deposit and migrate to the surface in a reasonable length of time. If 5 half-life intervals (20 days) is a reasonable length of time, about 1.5% of the ^{222}Rn will be left. Thus, there must be a path of sufficient permeability through the overlying sediments to allow the gas to reach the surface in twenty days. This method was used successfully in the Powder River Basin of Wyoming. An alternative is that the gas is carried by water or methane to the surface.

2.6 Optical Systems

2.6.1 UV Methods

Photoluminescence results from excitation of orbital electrons by absorption of electromagnetic radiation (i.e., light). This phenomenon can be useful to identify those minerals which exhibit one or more of the several forms of luminescence. Some of the more valuable minerals can be identified in the field in raw or dispersed form using these techniques.

Several forms of luminescence occur, such as *fluorescence* and *phosphorescence*. Luminescence is caused by the interaction of orbital electrons with electromagnetic photons. The electrons can absorb the energy from impinging photons and are lifted from orbit. After a period of time, the excited electrons will release their excess energy and drop back into their proper orbit. If the energy is in the form of secondary radiation, the phenomenon is known as photoluminescence. The secondary radiation is in the form of electromagnetic radiation and may be visible light. If the release is immediate or within a small fraction of a second, it is called fluorescence. If the electron wanders for more than that brief time, the emission is called phosphorescence. Fluorescence ceases when the exciting radiation ceases. Phosphorescence continues for an appreciable period of time after the exciting radiation ceases. Luminescent radiation has, generally, a longer wavelength than the exciting radiation. If the emitted radiation contains the same wavelengths as the source, the luminescence is called *resonant radiation*.

UV light is used as the exciting radiation in field equipment because of its convenience and relative safety. However, other types of radiation can be used: X-rays, gamma rays, visible light, etc. Visible light is not used in field equipment because its energy content is lower than UV, X-ray, and gamma radiation. Thus, it is less likely to cause excitation. Also, because our eyes respond readily to visible light, the luminescent light is masked by visible light as a source. The reflected visible light can be many times stronger than the luminescent light.

Luminescence in many minerals is due to included impurities in minerals. These are called *activating agents*. For example, calcite normally does not fluoresce. A trace amount of manganese impurity, however, will result in visible fluorescence. The activating manganese may only be 1 to 5% of the total (3.5% results in the most brilliant emission). Sodium iodide is a clear, nonfluorescing crystal. For use as a gamma ray detector, however, it is activated with trace amounts of thallium.

UV methods are mostly restricted to the laboratory. Field equipment, however, is readily available and used extensively in the petroleum business and tungsten mining. These units use UV radiation and generally use filters or other devices to allow "short" and "long" wavelength UV examination.

There are limitations to the use of UV for identifying minerals, as only a few of the commercially important minerals fluoresce. However, the simplicity and expedience of the method has made fluorochemistry a valuable tool in petroleum, uranium, mercury, and tungsten exploration efforts. Mining, sorting, grading, and milling of valuable minerals also benefit from the use of this technique. Lists of some of the minerals responding to fluorescent methods and their fluorescence colors are shown in Tables 2.1 and 2.2.

TABLE 2.1

A List of Some Minerals, their Formuli, and their Fluorescence Colors

Mineral	Formula	Usual Fluorescence Color
Agate	SiO_2	Green
Albite	$Na_2 \cdot Al_2O_3 \cdot 6SiO_2$	Bright green
Amber	Organic	Yellow (Texas), yellow-green (Prussia)
Aragonite	$CaCO_3$	Green, blue-white, yellow
Barite	$BaSO_4$	Yellow-white
Benitoite	$BaO \cdot TiO_2 \cdot 3SiO_2$	Deep blue
Calcite	$CaCO_3$	Red
Chalcedony	SiO_2	Green
Fluorite	CaF_2	Yellow, bright blue
Hyalite	SiO_2	Green
Hydrozincite	$ZnCO_3 \cdot 2Zn(OH)_2$	Soft blue
Mangan-apatite	$9(Ca,Mn)O \cdot 3P_2O_5 \cdot Ca(OH,F)_2$	Bright orange
Mercury	Hg	Green
Petroleum	Organic	Yellow, varies with
Powellite	$Ca(Mo,W)O_6$	Yellow, greenish
Scheelite	$CaWO_4$	Blue to white, brown
Selenite	$CaSO_4 \cdot 2H_2O$	Varies
Sphalerite	ZnS	Bright orange
Uranium	U**	See Table 22.5
Wernerite	Ca,Al,SO_4	Orange
Willemite	Zn_2SiO_4	Green

Chemically active wavelengths of UV are chiefly between 3000 and 4000 Å. For most fluoro-chemical purposes, they are classified as *long* (near 4 kÅ) and *short* (near 3 kÅ). Sources are usually quartz tungsten lamps with a filter. The better and newer ones use a mercury-gas discharge tube (containing argon,

TABLE 2.2

An Additional List of Some Minerals and their Fluorescence Colors

Uranium Mineral	Fluorescence Color
Autunite	Yellow-green
Beta-uranopilite	Yellow-green
Beta-uranotil	Yellowish
Chalcolite	Yellow-green
Gummite	Violet
Johannite	Yellow-green
Meta-torbernite	Yellowish-blue
Schroeckingerite	Green
Torbernite	Yellow-green
Uranocercite	Yellow-green
Uraniferous hyalite	Yellow-green
Uranophane	Yellow-green
Uranopilite	Yellow-green
Uranopathite	Yellow-green
Uranothallite	Green
Uranotil	Yellowish
Zippeite	Yellowish

neon, helium, and mercury). The latter emit primarily 2540 Å radiation (see Figure 2.13). Some of the newer tubes apparently have an interior coating to broaden the emitted UV spectrum.

Tungsten lamps must be heavily filtered to remove the visible light in order to allow the faint fluorescence to be seen. However, these are not the best sources, even with good filters. Electrical sparks between iron electrodes are rich in the 4270 to 2100 Å range, and are also rich in the visible portion and must be heavily filtered.

Luminescence, especially fluorescence, is valuable in mining and ore sorting of valuable minerals, such as tungsten, mercury compounds, and some gem stones. Many examples of fluorescence, among normally nonfluorescent materials, occur in minerals from specific locations. Such examples are amber in coal from Texas (yellow) and Prussia (yellow-green), calcite from New Jersey (red) and from Texas (pink or blue), and gypsum from the Southwest U.S. (green) and from Grand Rapids, MI (deep green). More complete lists can be found in the texts on mineralogy.

2.6.2 Visual Images

2.6.2.1 Video Logging

Video logging makes a visual image of the wall of a borehole as a function of hole depth. It is especially valuable for routinely examining monitor boreholes on landfill projects, gas wells, and determining casing damage wherever a cased or open borehole can be air-filled. Video logging uses a video

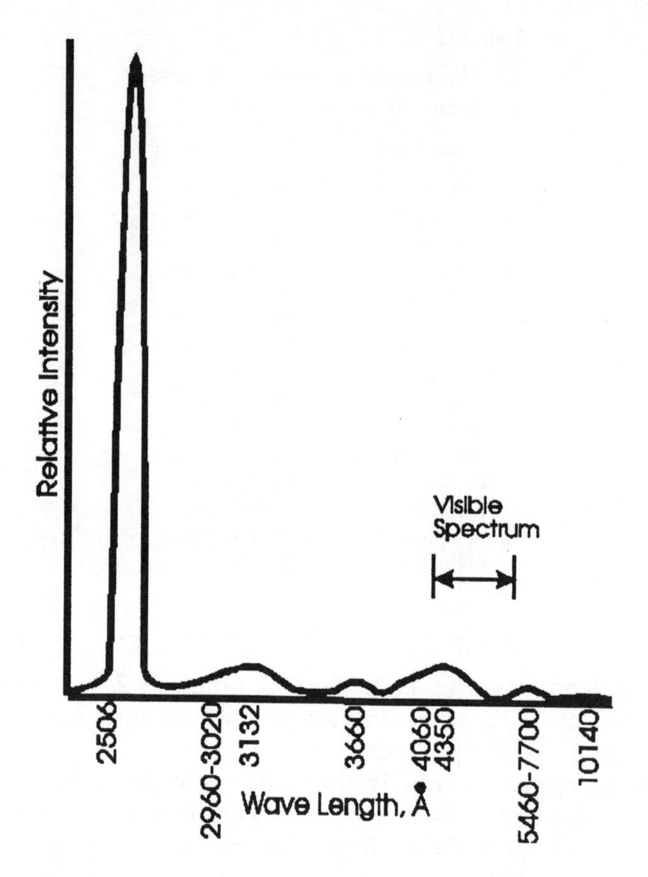

FIGURE 2.13
The spectrum of an arc discharge for a UV source.

camera downhole and digital data transmission. Analog versions are available, but the band-width of the logging cable presents some difficulty in obtaining satisfactory results. The downhole equipment is available in several sizes and designs. Recordings can only be made in gas- or air-filled boreholes. Limited success has been reported in logging in clear water-filled boreholes. Sizes are available for as small as 4 3/4 in. (12 cm) casing.

One form of the video logging downhole device uses a small video camera mounted vertically, pointed downward. A conical mirror is mounted below the camera, with the apex of the mirror at the lens center. High intensity krypton lights illuminate the borehole wall adjacent to the mirror resulting in a radial image that can be monitored on a screen in a logging truck. The image appears to move radially outward as the sonde is retrieved upward. The images are recorded on video tape.

2.6.2.2 Remote Sensing

Remote sensing results in visual images, of photographic type and quality, of the surface of the earth. This has been included to remind the reader that this valuable exploration tool is readily available from the U.S. government. The term *remote sensing* actually applies to any form of instrumentation where the detector is physically remote from the readout. Modern usage, however, has applied the term to only the methods employing aircraft and satellites and photographic methods.

Much progress has been made in the quality and scope of remote sensing in the last forty years. Images are gathered at many different wavelengths over the electromagnetic spectrum and at altitudes of a few feet to hundreds of miles. Imaging in the visual and infrared ranges is especially valuable. A variety of sensors, scanners, and cameras are routinely used. Sonar imaging should also be included as remote sensing. Platforms include aircraft, satellites, and ships.

3

Nuclear Methods

3.1 Introduction

Neutron methods have been covered in Chapter 8 of *Introduction to Formation Evaluation*, and Chapters 9 and 10 of *Standard Methods*. The neutron porosity system is presently used in nonhydrocarbon applications as a porosity measurement in lieu of the resistivity log in sediments in cased holes. Neutron activation elemental analysis also enjoys some use, but is not common in nonhydrocarbon applications.

The neutron is a neutral nuclear particle with a mass approximately equal to that of a hydrogen atom. Because of its neutral charge, the neutron is quite penetrating and loses energy to other material, chiefly by collision with the nuclei of the material. The most efficient energy transfer is to another particle of the same mass. Therefore, the neutron loses most of its energy during transit through a material to the hydrogen of that material. Thus, it is an excellent hydrogen detector.

Because most of the hydrogen in a sediment is in the water and hydrocarbon content of the pore spaces, neutron systems can be excellent porosity measurers. This accounts for their use and popularity in the petroleum business. They can also be used this way in nonhydrocarbon work.

The thermal neutron capture micro-cross section, σ, varies widely for various elements (see Table 3.1). It varies from 3.4 millibarns (mb) in carbon to 49 kilobarns (kb) in gadolinium. This contrast means that there is an excellent potential for mineral identification using the neutron system in hard rock environments. If we consider the macro-cross section (Σ, b/cm^3) of the rock material, excluding gases and liquids, we find that Σ varies from 0.38 b/cm^3 for graphite to 12.8 kb/cm^3 for kernite (see Table 3.2). Refer also to *Introduction to Geophysical Formation Evaluation* and *Standard Methods of Geophysical Formation Evaluation.*

Thus, if the neutron cross section were cross-plotted against, for example, the bulk density or the magnetic susceptibility, useful mineral identification could be obtained. Carried further, combinations of Σ, acoustic travel time, t, and magnetic susceptibility could be combined in an MN-type cross plot to identify specific mineral groups.

TABLE 3.1

The Capture Cross Sections of Several Elements to Thermal Neutrons

A	Element	Symbol	Z	S Barns	A	Element	Symbol	Z	S Barns
1.	Hydrogen	H	1.008	0.33	49.	Indium	In	114.82	19450.
3.	Lithium	Li	6.939	0.0376	50.	Tin	Sn	118.69	0.63
4.	Beryllium	Be	9.013	0.0092	51.	Antimony	Sb	121.75	5
5.	Boron	B	10.811	758.86	52.	Tellurium	Te	127.60	4.7
6.	Carbon	C	12.011	0.0034	53.	Iodine	I	126.904	6.2
7.	Nitrogen	N	14.007	0.075	54.	Xenon	Xe	131.30	24.5
8.	Oxygen	O	15.999	0.178	55.	Cesium	Cs	132.905	30.0
9.	Fluorine	F	18.998	0.0098	56.	Barium	Ba	137.33	1.2
11.	Sodium	Na	22.990	0.534	57.	Lanthanum	La	138.905	8.9
12.	Magnesium	Mg	24.305	0.064	58.	Cerium	Ce	140.12	0.73
13.	Aluminum	Al	26.981	0.232	59.	Praseodium	Pr	140.907	11.5
14.	Silicon	Si	28.086	0.160	60.	Neodymium	Nd	144.24	49
15.	Phosphorous	P	30.974	0.190	62.	Samarium	Sm	150.4	5820
16.	Sulfur	S	32.065	0.49	63.	Europium	Eu	151.96	4100
17.	Chlorine	Cl	35.453	33.44	64.	Gadolinium	Gd	157.25	49000
19.	Potassium	K	39.098	2.14	65.	Terbium	Tb	158.924	30
20.	Calcium	Ca	40.080	0.44	66.	Dysprosium	Dy	162.50	930
21.	Scandium	Sc	44.956	25.000	67.	Holium	Ho	164.930	65
22.	Titanium	Ti	47.90	6.1	68.	Erbium	Er	167.26	160
23.	Vanadium	V	50.941	5.06	69.	Thulium	Tm	168.934	115
24.	Chromium	Cr	52.996	3.1	70.	Ytterbium	Yb	173.04	37
25.	Manganese	Mn	54.938	13.3	71.	Lutetium	Lu	174.971	75
26.	Iron	Fe	55.847	2.56	72.	Hafnium	Hf	178.49	103
27.	Cobalt	Co	58.933	37.5	73.	Tantalum	Ta	180.948	22
28.	Nickel	Ni	58.71	4.54	74.	Tungsten	W	183.85	18.5
29.	Copper	Cu	63.546	3.8	75.	Rhenium	Re	186.2	85
30.	Zinc	Zn	65.38	1.1	76.	Osmium	Os	190.2	15.3
31.	Gallium	Ga	69.735	3.1	77.	Iridium	Ir	192.2	425
32.	Germanium	Ge	72.59	2.3	78.	Platinum	Pt	195.09	9.
33.	Asenic	As	74.922	4.3	79.	Gold	Au	196.967	98.8
34.	Selenium	Se	78.96	12.2	80.	Mercury	Hg	200.59	375.
35.	Bromine	Br	79.904	6.8	81.	Thallium	Tl	204.37	3.4
37.	Rubidium	Rb	85.467	0.5	82.	Lead	Pb	207.2	0.18
38.	Strontium	Sr	87.62	1.21	83.	Bismuth	Bi	208.980	0.034
39.	Yttrium	Y	88.906	1.3	84.	Polonium	Po	209	0.003
40.	Zirconium	Zr	91.22	0.182	86.	Radon	Rn	222	0.72
41.	Niobium	Nb	92.906	1.15	88.	Radium	Ra	226.025	20.
42.	Molybdenum	Mo	95.94	2.65	89.	Actinium	Ac	227	510.
44.	Rhuthenium	Ru	101.07	3.0	90.	Thorium	Th	232.038	7.4
45.	Rhodium	Rh	102.905	150	91.	Protactinium	Pa	231.036	260.
46.	Paladium	Pd	106.4	6	92.	Uranium	U	238.029	7.59
47.	Silver	Ag	107.868	63.8					
48.	Cadmium	Cd	112.41	2450					
49.	Indium	In	114.82	19450.					

Note: Σ is in b/cm^3.

TABLE 3.2

The Capture Cross Sections of Various Compounds to Thermal Neutrons

Mineral	Formula	Σ
Albite	$NaAlSi_3O_8$	6.77
Aluminum	Al	13.99
Anhydrite	$CaSO_4$	12.30
Aragonite	$CaCO_3$	8.12
Arsenopyrite	FeAsS	165.22
Barite	$BaSO_4$	19.40
Biotite	$H_2K(Mg,Fe_3Al(SiO_4)_3$	25.20
Bornite	Cu_5FeS_4	145.63
Calcite	$CaCO_3$	7.48
Carbon dioxide	CO_2	
Carnalite	$KMgCl_3 \cdot 6H_2O$	370.92
Carnotite	$K_2O \cdot 2UO_3 \cdot 2H_2O$	56.21
Cement		~13
Chalcopyrite	$CuFeS_2$	80.95
Chalcosite	Cu_2S	173.56
Chlorite	$(Mg,Al,Fe)_{12}(Si,Al)_8O_{20}(OH)_{16}$	17.56
Chromite	$FeCr_2O_4$	102.20
Cinnabar	HgS	7981.16
Coal, bituminous		1.54
Coal, anthracite		1.08
Cobaltite	CoAsS	936.73
Corundum	Al_2O_3	11.04
Diabase		17.12
Diorite		14.33
Dolomite	$CaMg(CO_3)_2$	4.78
Dunite		17.03
Gabbro		21.47
Galena	PbS	12.47
Glauconite	$KMg(FeAl)(SiO_3)_6 \cdot 3H_2O$	16.80
Granite		11.62
Granodiorite		11.33
Graphite	C	0.38
Gypsum	$CaSO_4 \cdot 2H_2O$	19.40
Halite	NaCl	752.36
Helium	He, STP	0.00
Hemitite	Fe_2O_3	100.47
Hydrogen sulfide	H_2S, STP	
Illite	$KAl_5Si_7O_{20}(OH)_4$	39.90
Illmanite	$FeTiO_3$	158.23
Iron	Fe	214.90
Kainite	$MgSO_4 \cdot KCl \cdot 3H_2O$	196.13
Kaolinite	$(OH)_8Al_4Si_4O_{10}$	13.06
Kernite	$Na_2B_4O_7 \cdot 4H_2O$	12,793.69
Kieserite	$MgSO_4 \cdot H_2O$	12.77
Langbeinite	$K_2Mg_2(SO_4)_3$	78.87
Lead	Pb	5.61
Limestone		8.72
Limonite	$2Fe_2O_3 \cdot 3H_2O$	74.10
Magnesite	$MgCO_3$	1.48

TABLE 3.2 (continued)

The Capture Cross Sections of Various Compounds to Thermal Neutrons

Mineral	**Formula**	Σ
Magnetite	Fe_3O_4	112.10
Montmorillonite	$(OH)_4Si_8Al_4O_{20}\cdot nH_2O$	8.10
Muscovite	$KAl_2(AlSi_3)O_{10}(OH)_2$	17.30
Nitrogen	N_2, STP	
Norite		12.88
Olivine	$(Mg,Fe)_2SiO_4$	31.74
Orthoclase feldspar	$KAlSi_3O_8$	16.00
Oxygen	O_2, STP	1×10^{-5}
Plagioclase feldspar	$xNaAlSi_2O_8,yCaAl_2Si_2O_8$	6.99
Polyhalite	$2CaSO_4\cdot MgSO_4\cdot K_2SO_4\cdot 2HO$	21.00
Potash	$KCO_3\cdot 2H_2O$	39.70
Pyrite	FeS_2	89.06
Pyrrhotite	Fe_5S_6	90.52
Quartz	SiO_2	4.36
Rhodocrosite	$MnCO_3$	287.93
Rutile	TiO_2	202.75
Sandstone		
Serpentine	$(Mg_3Si_2)_5(OH)_4$	8.80
Siderite	Fe_2CO_3	68.81
Sphalerite	ZnS	38.33
Stibnite	Sb_2S_3	17.82
Sulfur, monoclinic	S	18.05
Sulfur orthorhombic	S	19.06
Sylvite	KCl	570.68
Synite		16.43
Trona	$Na_2CO_3HNaCO_3\cdot 2H_2O$	16.21
Uraninite	UO_2	49.69
Water, (1.5×10^5 ppm NaCl)		78.75
Water, (2.0×10^5 ppm NaCl)		100.08
Water,(2.5×10^5 ppm NaCl)		122.55
Water,(5.0×10^4 ppm NaCl)		39.02
Water,(1.0×10^5 ppm NaCl)		58.69
Water, (3.0×10^5 ppm NaCl)		146.22
Water, (3.0×10^4 ppm NaCl)		32.56
Water, pure, STP	H_2O	22.08
Wulfenite	$PbMoO_4$	32.50
Zircon	$ZrSiO_4$	5.42

Note: $\Sigma = b/cm^3$.

Neutrons react rapidly with a nearby atom's nuclei when the neutron has lost most of its energy; that is, when its energy is at an epithermal or thermal level. At these levels, especially at the thermal level, the neutron can be captured and the excess energy given off as a characteristic gamma ray. The element of the capturing atom can be identified by the wavelengths of these captured gamma rays. This type of equipment is available as downhole systems of several forms, sizes, and sophistication from both the oil and the mineral logging contractors. It is also available at many of the core and mineralogy laboratories.

Princeton Gamma Tech (PGT) of Princeton, NJ, experimented briefly with their ultrapure germanium detector spectral natural gamma ray system and a 5×10^6 n/s AmBe neutron source to locate gold in a borehole (direct communications, around 1980). The experiment was considered successful in that the line of gamma ray emission from gold was detectable. A stronger source appeared to be needed.

These uses of neutron systems have been discussed in earlier volumes and will not be repeated here. As far as this author is aware, there are no further uses for the neutron systems, except for forensic uses. They are frequently used by forensic laboratories for sample analysis. These systems are also a potentially valuable tool for security uses, explosive detection, medical analysis, and investigation of art work.

3.2 Radon Surveys

Radon detection is also discussed in Chapter 2. Radon decays by the emission of an energetic particle, the alpha (α) particle. The α particle is the nucleus of the helium (He^4) nucleus.

The free radon content of the soil varies widely. It depends upon a number of factors including the air temperature, changes of temperature, ambient air pressures, pressure changes, the presence of disseminated radioactive minerals, the amount and distribution of clay minerals, and the presence or lack of anomalous uranium deposits. It is not possible to assign a unique radon value to an area. Probable values, comparable values, and trends, however, can successfully be used.

Because radon is a noble gas (chemically inert), it can migrate freely through porous rock and soil without reacting with local materials. It occurs as ^{222}Rn from ^{238}U, ^{220}Rn from ^{220}Th, and ^{219}Rn from ^{235}U. It diffuses through the enclosing or parent material after forming by decay. It is, however, unlikely to escape from the mineral grain unless the parent radium is close to the grain surface. This can occur if the grain is small or if the grain has a thin crust of radium from hydrothermal or weathering alteration.

After the radon escapes from the mineral, it will diffuse through the ground air and water and will dissolve in the water to a small extent. With a half-life of nearly 4 days, ^{222}Rn can diffuse the greatest distance from its source. It is unlikely that ^{220}Rn, with a half-life of 52 sec, nor ^{219}Rn, with a half-life of 3.9 sec, will be found outside the immediate vicinity of their sources.

Arid soils have almost complete continuity between ground and atmospheric air. Thus, low atmospheric pressures and strong winds tend to draw the soil gases upward and reduce the radon concentrations in the soil. Calm, high pressure conditions and wet topsoil tend to restrict the flow from the soil and allow radon to concentrate in depth, in the soil. Humic topsoil can seal off the top thin layer and result in a concentration of radon near the surface.

The diffusion length of ^{222}Rn in water is severely limited. Groundwater flow, even in the neighborhood of springs, is low; on the order of 1 ft (30 cm/day). Studies in Canada (Dyck and Smith, 1968) show that most of the radon found in surface waters is derived from radium absorbed in bottom sediments or from springs feeding lakes and streams. Complete saturation of the soil (with water) limits the diffusion length of ^{222}Rn to about 10 cm (4 in.) in water. Therefore, any significant migration of ^{222}Rn will depend upon the flow of the water bearing it. Figures 3.1 through 3.4 show the effects of atmospheric conditions upon soil air and humic conditions, respectively.

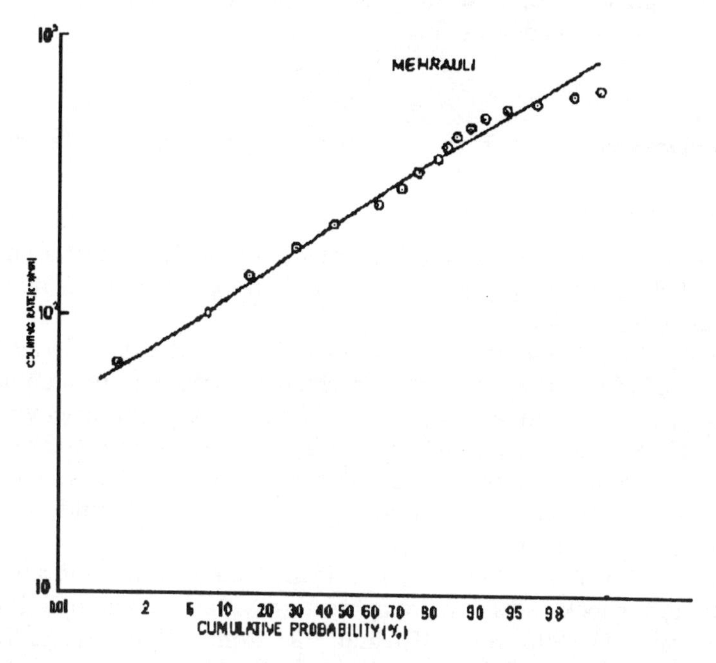

FIGURE 3.1
The radon detection over a known barren field.

Radon surveys of soil air have their greatest potential in areas where uranium deposits are known or suspected to exist.

The use of radon measurement for uranium exploration was suggested by Behounek in 1927. He had studied the anomalous radioactivity he had found in springs, soils, and in the atmosphere around deposits in Joachimthal, Czechoslovakia. Field prospecting was attempted by the Soviets in the early 1930s using soil air. Further studies were made when uranium exploration became more important after 1945, but was not widely used prior to 1960.

In one series of tests (Bhatnagar, 1973), multiple boreholes were used in each of 14 locations. Some of the locations were known to have no

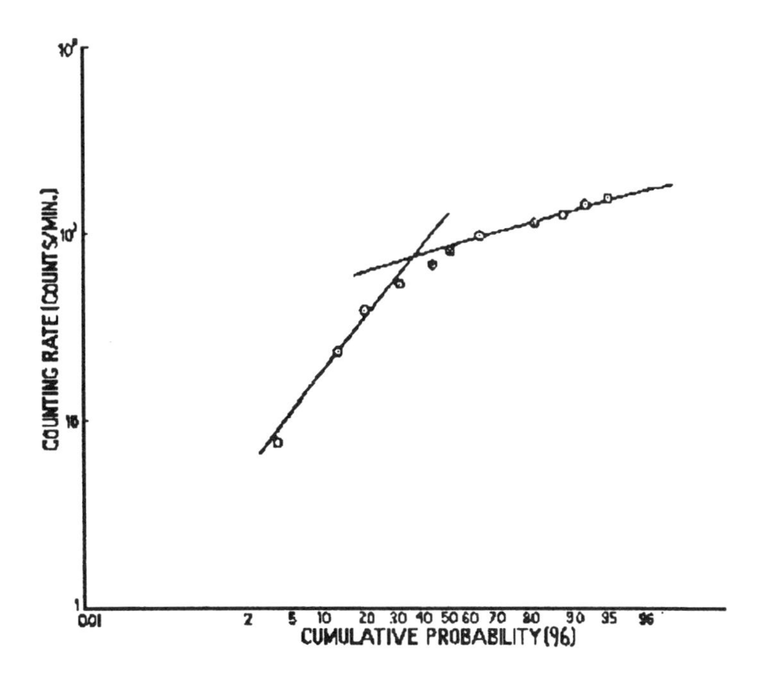

FIGURE 3.2
The radon measurement over a known mineralized field.

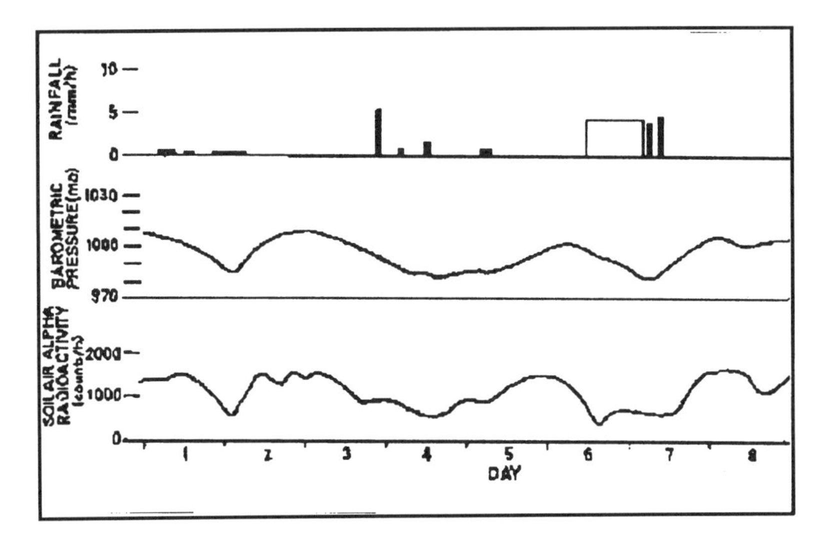

FIGURE 3.3
The effect upon radon evolution from rainfall and barometric pressure.

mineralization. Others were known to have above background uranium mineral deposits. In each borehole, a phosphor-coated chamber was lowered into the air-filled borehole. Borehole air was circulated through

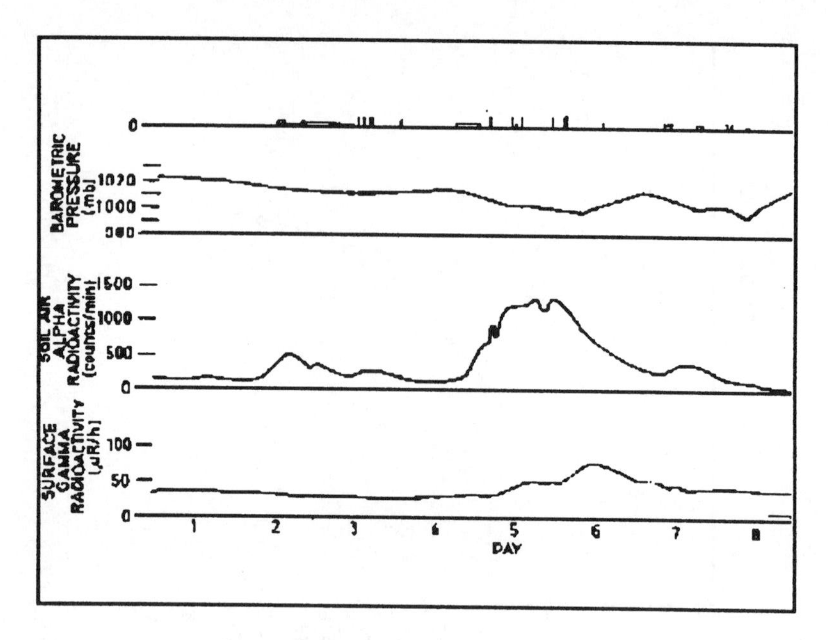

FIGURE 3.4
The effect of barometric pressure upon the soil radon content.

the chamber. The chamber was then sealed and brought to the surface. The radon radioactivity of the contents was measured. Log-normal plots were made for each location.

The plots of the barren area boreholes showed a simple, single distribution. Those from mineralized locations tended to show bimodal distributions.

In a further test, soil samples were taken from around a borehole in a barren area. A single distribution was found for the radon contents of the samples. A radium chloride source was lowered into the borehole. After several weeks, soil samples within 1 foot (30 cm) of the borehole began to show the bimodal distribution. Those farther out remained a simple single trend.

3.2.1 Alpha Cups

During the decade of the 1960s, a radon survey method was successfully used for uranium exploration which proved quite valuable and relatively inexpensive. The method used radon or alpha cups. This method makes use of the radon (^{222}Rn) which is one of the important daughters of the uranium (^{238}U) decay. Because radon is a gas, some part of it may migrate toward the surface above a uranium or thorium deposit. Since ^{222}Rn has a half-life of about four days, there is a substantial probability that some will reach the surface.

^{222}Rn is an alpha particle emitter when it decays to astatine. A photographic film, in the path of an alpha particle, will show a track marking the path of the particle. After the film has been developed, the number of tracks can be counted. This, and more sophisticated techniques are common in the radiation protection badge industry.

In the alpha cup method, the film is enclosed in a thin, light-tight cover and fastened in the bottom of a cup. Hundreds or thousands of these cups are placed, inverted, in a grid, in shallow depressions and covered with earth and left for several days. After development and counting, the values are plotted on a map and the isorad lines drawn. The isorads will tend to encircle any radioactive deposit. The intensity will be a function of the mineral amount, type, and overburden type.

The alpha cup method depends upon some of the ^{222}Rn being able to escape the deposit and migrate to the surface in a reasonable length of time. If 5 half-life intervals (20 days) is a reasonable length of time, about 1.5% of the ^{222}Rn will be left. Thus, there must be a path of sufficient permeability through the overlying sediments to allow the gas to reach the surface in 20 days. An alternate is that the gas be carried by water or methane to the surface. This method was used successfully in the Powder River Basin of Wyoming.

4

Magnetic and Electromagnetic Methods

4.1 Introduction

There has been some activity lately in downhole magnetic methods. This is following the trend of the long-useful airborne and surface methods. These latter fields have been useful and active for many years, and many good methods are potentially available.

It is probable that downhole magnetic methods will become more important in the immediate future, as their usefulness is learned and their value is recognized. Work has been done in the past by the U.S. Geological Survey and the U.S. Department of Energy. Instrumentation problems, however, have presented obstacles. These problems have mainly been temperature drift and interpretation. The contractors who build their own equipment have discussed these problems and are gradually offering viable services in the realm of magnetic measurements.

Surface and airborne magnetic measurements should certainly be considered in combination with other geophysical and geological techniques for wide-area exploration. These methods are established, and the instrument environment lends itself to the control of the known problems. These methods are especially valuable when combined with surface geology and borehole geophysics.

The several magnetic methods are surface and airborne magnetic field strength, sample magnetic susceptibility, and downhole magnetic susceptibility. Closely related are some of the borehole deviation measurements, casing collar location, detection of "fish," and induction logging systems. These methods will be discussed.

4.2 Magnetic Relationships

The magnetic flux density in empty space is related to the magnetic permeability of free space and the magnetic intensity or field strength:

$$B_0 = \mu_0 H \tag{4.1}$$

where

B_0 = the magnetic flux density in free space (Wb/m^2)
H = the magnetic intensity or field strength
μ_o = the magnetic permeability of free space, in 4×10^{-7} H/m

Then,

$$\mu = \mu_o + x \tag{4.2}$$

where x = the magnetic susceptibility.
Thus,

$$B = \mu H \tag{4.3}$$

where μ = the total magnetic permeability. And,

$$K_m = \frac{\mu}{\mu_0} = 1 + \frac{\chi}{\mu_0} \tag{4.4}$$

where K_m = the relative magnetic permeability. Also, for a toroid,

$$B = \mu \frac{Ni}{L} \tag{4.5}$$

where

N = the number of turns of the coil
i = the electrical current in the coil
L = the length of the coil

If we examine a long coil (compared to its diameter) having a core of magnetic material, the field far from each end will be uniform. Then, combining Equations 4.3 and 4.5:

$$H = \frac{Ni}{L} \tag{4.6}$$

The magnetic moment, g, is the torque apparent on a magnetic dipole when it is 90 degrees to the existing magnetic field:

$$g = \mu_0 \left(\frac{Ni}{L} \right) = \chi H \tag{4.7}$$

Therefore,

$$B = \mu_0 H + g \tag{4.8}$$

Returning to Equation 4.3, in the absence of magnetic material in the coil core and in free space, we have

$$B_0 = \mu_0 H \tag{4.1}$$

Magnetic susceptibility is usually measured by putting an alternating magnetic field into the formation and measuring the magnetic field strength. This type of instrument is very similar to an induction logging instrument. It measures the out-of-phase component of the field, rather than the in-phase (as does the induction log). In fact, the new array induction of Schlumberger measures the susceptibility component and uses it to make corrections to the in-phase signal (the conductivity component).

Most magnetic susceptibility measurements are made on surface samples in the laboratory. In this case, the temperature problems can be easily handled. Also, the measurement instruments are available commercially.

The important features of the magnetic susceptibility methods are:

1. The magnetic susceptibility (or range of susceptibilities) of a mineral is characteristic of that mineral.
2. Iron, which, in some forms, exhibits large susceptibility values, is frequently associated with valuable sedimentary mineral deposits.
3. Changes in the susceptibilities of many compounds take place when their redox states are changed.

These methods should be examined closely. They may be valuable methods in both sediments and in hard rock environments. Table 4.1 is a partial list of the magnetic susceptibilities of some minerals.

4.3 Surface Measurements

Magnetic field strength measurements are commonly made to display a pattern of isomagnetic lines. The shape and concentration of these lines can be interpreted in terms of the presence of bodies of surface and subsurface materials that distort magnetic lines. These measurements are commonly made with aircraft, boats, or automobiles traveling in a predetermined grid, making measurements of the field strength. These measurements are then plotted as isomagnetic lines.

TABLE 4.1

A Partial List of Susceptibilities

Material	Temperature (°K)	k (10^{-6} cgs)
Aluminum	293	+16.5
Barium	293	+20.6
Carbon	293	-5.9
Chromium	273	+180.0
Gold	293	-61.0
Iron, FeO	293	-7200
Iron, Fe_2O_3	1033	+3586
Lead oxide, PbO	273	-42.0
Lead sulfide, PbS	273	-84.0
Magnesium carbonate, $MgCO_3 \cdot 3H_2O$	273	-72.2
Nickle oxide, NiO	293	+660
Silicon oxide, SiO_2	293	-29.6
Sodium chloride, NaCl	293	-30.3
Uranium, U, α	78	+395
Uranium, U, α	298	+409
Uranium, U, α	623	+440
Water, H_2O, (liquid)	373	-13.1
Water, H_2O, (liquid)	293	-13.0
Water, H_2O, (liquid)	273	-12.9
Water, H_2O, (gas)	273	-12.6
Water, H_2O, (gas)	223	-12.3
Zinc oxide, ZnO	293	-63.0
Zinc sulfide, ZnS	293	-25.0

From Weast, R., Ed., *Handbook of Chemistry and Physics*, 61st ed., CRC Press, Boca Raton, 1980.

Presently, magnetic susceptibility logging is not a widely used technique. It does have considerable promise, however, especially in mineral exploration. One contractor is offering this service as a routine part of the induction logging technique. Much of the problem, in the past, has been temperature effects upon the instruments.

Susceptibility measurements offer a means of identifying certain ferromagnetic minerals that are typically associated with some types of mineral deposits. They are also useful in determining the state of some mineral compounds. Magnetic techniques are important to uranium and iron exploration.

4.3.1 Geophysical Measurements

The logging measurement is one of measuring an alternating current (AC) in an unbalanced Maxwell-Wien bridge. In operation, the magnetic coupling between the impedance bridge is null in the air before logging. The bridge and the formation surrounding the borehole drive the bridge circuit to a condition of imbalance. The AC voltage and the frequency across the bridge are held constant. The impedance of the inductive leg of the bridge changes, in the borehole, because of:

1. Changes in the external magnetic field "paths" that link to the iron core of the inductor.
2. The induced currents that circulate horizontally about the inductor, caused by the changing (alternating field) in the formation.

The changes in the field paths are proportional to the local susceptibility (k) of the formation. They are also manifested in the reactive (or quadrature) component of the impedance. It is the change of the bridge impedance that is proportional to the sample susceptibility and conductivity. This is the detection scheme used by the induction logging systems. It is an alternate method (the one used in induction logging equipment) used to detect the amplitude of the out-of-phase signal occurring in the receiver coils. The resistive (in-phase) and reactive (out-of-phase) components are separated by phase-sensitive detectors. The change in the resistive (or in-phase) component of the signal is proportional to the conductivity (C) of the formation material within the field. The susceptibility and conductivity are recorded individually. Ellis (1987), Tittman (1986), and Moran and Kunz (1962) offer useful explanations for the operation of magnetic induction geophysical instruments.

4.3.2 Application

The magnetic susceptibility log is not a new technique, although it has been neglected for years because of lack of interest. Because most of the obvious applications are in the mineral industry, there has not been much investigation of the technique because of lack of funding.

The susceptibility measurement has been mentioned (Ellis et al., 1987) as a possible prospecting tool because of a negative correlation with the occurrence of uranium. A regional study in sandstone also demonstrated a negative correlation. It has been suggested the alteration of magnetite (Fe_2O_3) and ilmenite ($FeTiO_3$) is the principle cause of the negative correlation with uranium mineralization.

A logging measurement is done rather quickly and with high sensitivity; 10 μcgs, or less, is fairly typical. Logging, however, frequently is hampered by signal drift caused by temperature and ambient pressure changes. These problems appear to have been eliminated or minimized with the induction of a log-associated system.

Thin bed response of the bridge-type earlier equipment is shown in Figure 4.1. These are experimentally determined responses measured in beds of sand of different thicknesses. The coil used was 39 cm (15 in.) in length. The measured response to bed thickness agrees well with the calculations for a 39-cm coil.

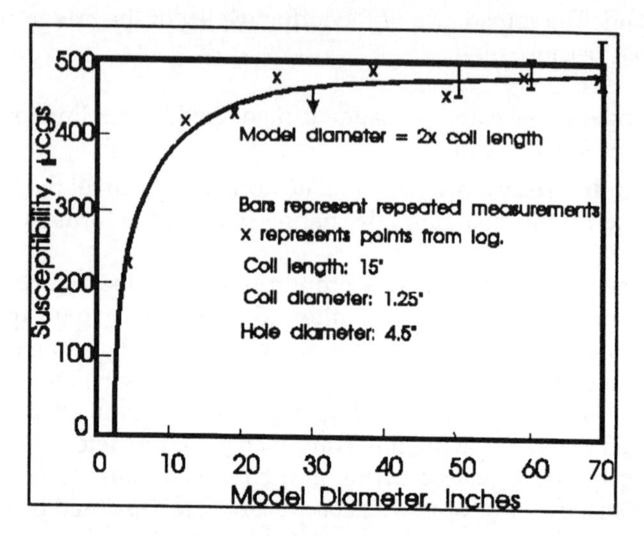

FIGURE 4.1
The response of a bridge-type measurement of magnetic susceptibility. (After Crone.)

The response shape associated with the induction equipment is not known, as the susceptibility signal is primarily used as a correction component.

The magnetic susceptibility log, of course, is not useful in iron-cased boreholes, but it can be used in plastic-cased and open boreholes.

Absolute calibration is difficult without uniform, homogeneous models of known magnetic susceptibility. The sensitivity of the measurement makes it difficult to construct suitable models of known magnetic susceptibility.

4.4 Electromagnetic Systems (EM)

4.4.1 Variable Frequency Systems

A number of surface electromagnetic (EM) systems that employ a frequency-domain system are commonly used. A sinusoidal signal is varied in frequency, and the amplitudes of the received signals are compared to the primary signal. The signal is a function of the conductivity of the earth.

4.4.2 Pulsed Systems

4.4.2.1 Principle

These methods are also called "transient electromagnetic techniques". A large coil, usually a single turn, is placed horizontally on the surface above the target zone. A large DC current is then flowed through the coil. This causes a magnetic field to grow and stabilize in the earth, below the coil. (There is, of course, a similar field above the coil in the air. This, however, does not enter the picture.)

When the magnetic field and resulting eddy currents have stabilized in the earth, the current is abruptly terminated. This induces a large, short current pulse in the conductive formation materials. The current is proportional to the conductivity of the formation material.

The induced current creates a secondary magnetic field that is proportional to the current (thus, to the formation conductivity). This magnetic field is detected by the surface coil and is read as a function of the formation resistivity. Thus, it is obvious that the EM principle is reminiscent of the induced polarization and the induction log principles. Figure 4.2 shows the theoretical currents and magnetic fields of the pulsed EM system. Newmont Mining uses a long wire transmitter pulsed with a square pulse. The same wire is the receiver. Veliken and Bulgakov (1967) used a single, large horizontal coil for the transmitter and receiver. It was later improved by investigators in Australia.

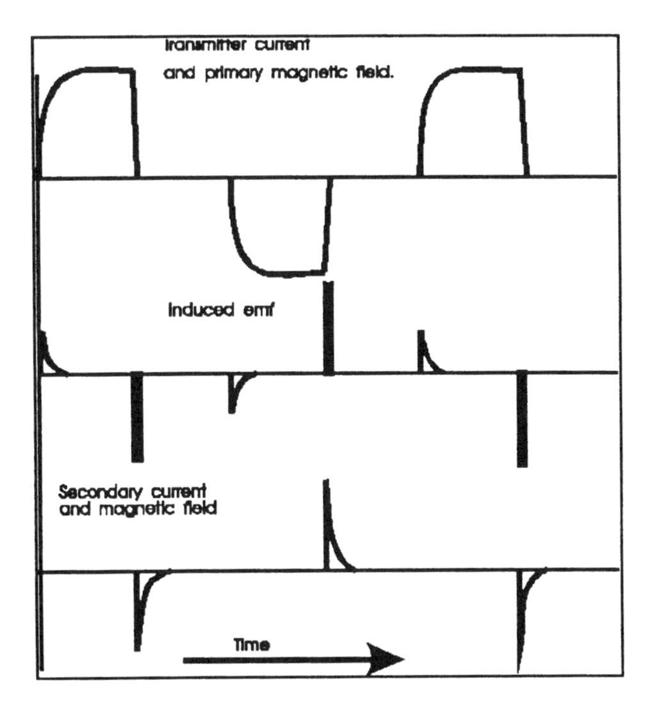

FIGURE 4.2
The theoretical currents and magnetic field of the pulsed EM.

Crone Geophysics and Newmont developed a pulsed system with the receiver synchronized by radio to examine, after the primary pulse, eight preselected time windows. This system was also modified for downhole detection with a surface transmitter.

Downhole pulsed EM probes have been patterned after the Doll/Schlumberger induction logging sonde. Ward and Harvey (1964) oriented a large transmitter coil on the surface, parallel to the axis of the borehole. A downhole coaxial receiver was used to reduce mutual coupling, which is a common fault with this type of system and results in a large, constant background signal. This system was very sensitive to borehole deviation, as might be expected.

The Crone system (PEM) uses a pulsed signal in a large horizontal coil. The signal is monitored by a small stationery coil to ensure it remains constant. A downhole coaxial coil measures the signal, after the termination of the primary pulse. This system is not sensitive to borehole deviation.

Geonics Ltd. uses a large coil for both the transmitter and the receiver. The received signal is sampled several times at frequent time intervals to determine its die-away character.

4.4.2.2 Advantages and Disadvantages of the Pulsed EM Techniques

It is often said that transient or pulsed EM measurements are essentially quadrature-phase determinations, because the measurement is made in the absence of a primary field. Although it is true that the targets that exhibit in-phase response only (such as nonconductive bodies with $\mu > \mu_0$) also yield no response after termination of the primary field, it can be shown that, in general, the transient method is responsive to both the in- and the quadrature-phase components of the target response.

It is also thought that since the transient and the multispectral responses are related through the Fourier transforms, the techniques are equivalent. While theoretically correct, this statement ignores the important influence of the major source of noise for each technique. Because of the noise influence, they are not practically equivalent, and transient techniques can offer a major advantage (McNeill, 1980).

The transmitter current wave form can be described by the Fourier sum of a fundamental and an infinite series of odd harmonics.

4.4.3 Electromagnetic Propagation

Oilfield geophysical logging contractors offer two downhole electromagnetic propagation systems which deserve mention. One of these is the Electromagnetic Propagation Time system (EPT), a shallow investigation system. The other is a Deeper Penetration Electromagnetic Propagation time device (DPT). The existent electromagnetic propagation time systems are specifically designed for petroleum saturation determinations. The EPT operates at 1 GHz and the DPT operates at 25 MHz. They essentially measure the dielectric permittivity, ε, and the electrical conductivity, C.

Hydrocarbon saturation determinations, with the traditional resistivity tools, are difficult to make when there is little resistivity contrast between formation water and the hydrocarbon. This occurs when the formation water is fresh (very low salinity) and its resistivity approaches that of the hydrocarbons.

EPT systems make use of the attenuating effect of the value of the dielectric constant upon the propagation of electromagnetic waves. The dielectric constant of a substance is the ratio of the capacitance of a capacitor using the substance as a dielectric between the plates to the capacitance of the same capacitor with only a vacuum between the plates.

The dielectric constant of water is high (78.58 at 77°F or 25°C and 34.5 at 392°F or 200°C). It is unaffected by the presence of any dissolved solids. The dielectric constants of hydrocarbons run from 1.6 to 2.5 over the same temperature range. Thus, these tools can be used with porosity values to determine hydrocarbon saturations in fresh water formations as an alternative to the resistivity/porosity combination.

Recent advances in instrument design, especially downhole, have greatly improved this method. Systems are still large and long, however. The use of the electromagnetic propagation time (EPT) systems is limited in nonhydrocarbon work because of poor interest in saturation and because of the sizes and lengths of the oilfield downhole tools. They do detect free hydrogenous materials (i.e., water), compared to bound water of clays. This might be of interest in water well production determinations. Table 4.2 shows the dielectric constants and propagation times for several sediments and fluids. Some references about the systems have been included in the bibliography.

TABLE 4.2

The Dielectric Constants for Several Sediments and Fluids

Material	Frequency	Dielectric Constant
Barium sulfate	10^8	11.4
Calcium carbonate	10^4	6.14
Calcium sulfate	10^4	5.66
Cupric oxide	10^8	18.1
Cupric sulfate·$5H_2O$	6×10^7	7.8
Dolomite	10^8	6.8-8
Ferrous oxide	10^8	14.2
Malachite (mean)	10^{12}	7.2
Quartz	3×10^7	4.27-4.34
Sulfur (mean)		4.0
Water	T = 25°C	78.54
Water	T = 200°C	34.5

From Weast, R., Ed., *Handbook of Chemistry and Physics*, 61st ed., CRC Press, Boca Raton, 1980.

4.5 Nuclear Magnetic Methods

Nuclear magnetic (nuclear magnetic resonance, NMR) logging may have some applications in nonhydrocarbon exploration. The present logging systems are long and large in diameter, because they are designed for oilfield conditions. The principle appears to be useful, however, especially when used in conjunction with porosity measurements. Laboratory methods are presently available. Figure 4.3 shows the assembly of an NMR probe.

This method determines the amount of free water (and/or hydrocarbon) within the sample, whether in the laboratory or downhole. It is not influenced by rock parameters, only the free hydrogenous fluids. This method ignores the bound water in clays and shales and the water of crystallization allowing for a good determination of effective porosity and of permeability.

This method aligns the free protons (hydrogen nuclei) with an intense magnetic field normal to the earth's magnetic field. When the instrument field is terminated sharply, the protons tend to return to their natural position by precessing around the earth's magnetic field lines. The resulting changing magnetic field generates an alternating current in a pickup coil. The maximum voltage of the signal is proportional to the number of participating hydrogen nuclei. The motion of the hydrogen nucleus is shown in Figure 4.4 and is described by Equations 4.5 and 4.6. Figure 4.5 shows the resulting signal.

4.6 Magnetic Susceptibility Logging

One of the common geophysical techniques for mapping is the measurement of magnetic susceptibility. This technique is widely used, especially for the search for iron-rich ores. The measurement of magnetic susceptibility is done from the air, on the surface, at sea, and from a borehole. The latter is not as common nor well known as the others.

The magnetic measurement techniques can measure magnetic susceptibility and/or residual permanent magnetism. Borehole measurements have been investigated by the U.S. Department of Energy, the Geological Survey of Canada, and by several corporations, two of which are Schlumberger Well Services, Inc. and Geonics, Ltd.

The investigations by the U.S. Department of Energy were conducted out of Grand Junction, CO, in the Powder River Basin (among other places), about 1970. They successfully used a two-coil sonde, but discovered some severe temperature drift problems. The experiments were reported in U.S. Department of Energy publications. The investigations by the Geological

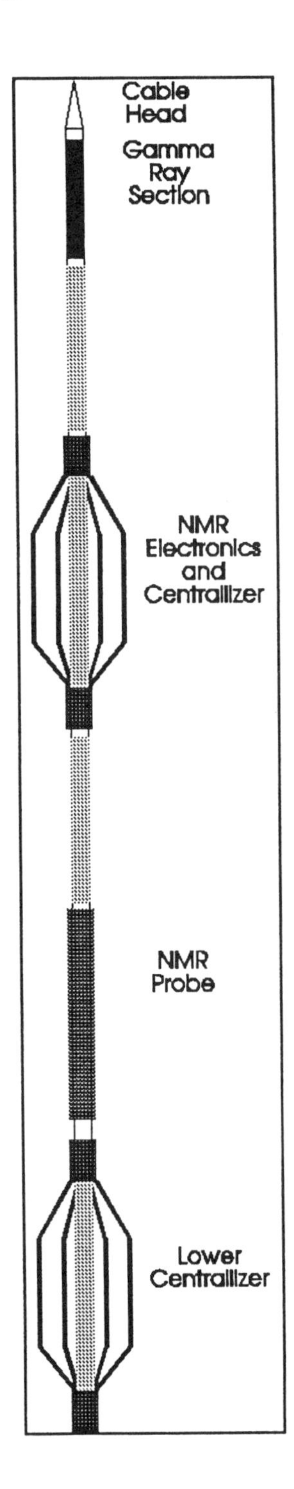

FIGURE 4.3
The NMR downhole tool.

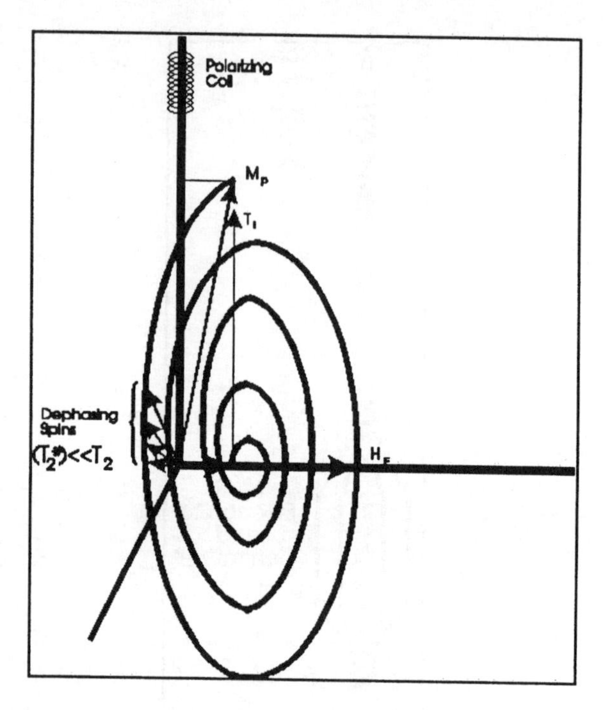

FIGURE 4.4
The motion of a hydrogen atom upon cancellation of a magnetic field. (Courtesy of Schlumberger Well Services, Inc.)

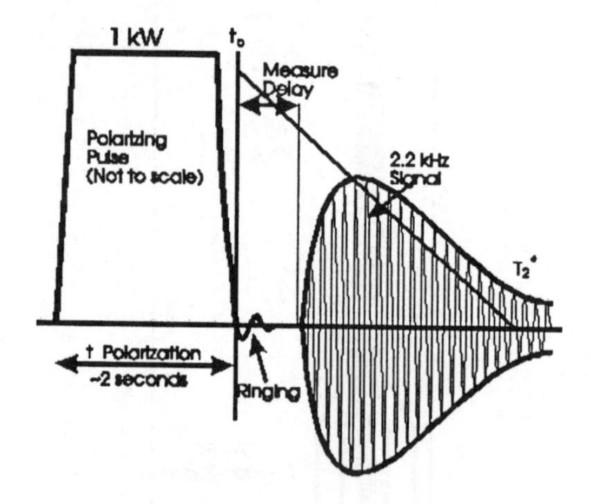

FIGURE 4.5
The signal resulting from the motion of the hydrogen atom. (Courtesy of Schlumberger Well Services, Inc.)

Survey of Canada and Geonics were reported publicly (McNeill et al., 1996). The investigations by Schlumberger have been numerous, as they involved work on the induction log, which they originated (Doll, 1949).

The induction log is included in a discussion of magnetic susceptibility because the common form of borehole magnetic susceptibility instrument is very similar to a simple, two-coil induction logging system. With this type of borehole instrument, the conductivity signal needed for oceanographic and petroleum use appears as the quadrature phase component of the received magnetic field. The magnetic susceptibility appears as the in-phase component. In oilfield instruments (the induction log family), the two components are separated by a phase-sensitive detector. This is because their use is in high conductivity sediments where the conductivity component is strong. In the Schlumberger "Phasor" Induction Log, the magnetic susceptibility component is actually used to correct some errors inherent in the conductivity signal (Schlumberger, c1993).

The U.S. Department of Energy (DOE) and the Geological Survey of Canada/Geonics tests both reported drift problems due to temperature changes, resulting in dimensional changes in the instrument coil spacing. Temperature changes are, of course, naturally present in boreholes. It is probable, too, that changes from the site of calibration to the borehole could also be troublesome.

4.6.1 Magnetic Materials

Relatively few magnetic materials have geologic significance. Iron ores, themselves, are important. The circumstances of their presence, however, can be quite diagnostic and important. Some magnetic materials are often co-deposited with industrially important minerals, such as uranium. The amounts and distributions of such magnetic materials within sedimentary and "hard" rocks can be quite informative.

Magnetite (Fe_3O_4) is, by far, the most important magnetic material. Pyrrhotite (Fe_7S_8), ilmenite (Fe_2TiO_3), and maghemite (γFe_2O_3), are also significant. Maghemite is a form of hematite (Grant et al, 1965). Note that *most* of the magnetically active forms of iron compounds are the reduced forms. The more oxidized forms tend to be less active magnetically.

Magnetite is dark, heavy, and extremely resistant to weathering (McNeill et al., 1996). This makes it useful as an indicator mineral because it survives many of the geological alteration processes. Magnetite sometimes shows natural magnetic polarity and was used earlier to attract iron. Slivers and elongated lumps were used as compasses. Because it is heavy, magnetite frequently occurs with placer gold deposits, resulting in the use of magnetometers to locate gold deposits. Magnetite and other reduced forms of iron are frequently associated with geochemical, sedimentary uranium deposits ("roll fronts"). The deposits probably occur because of the close proximity of the iron's respective electrochemical potentials and relative insolubilities ($Fe^{+2} = -0.41$ v, $U^{+4} = -0.60$ v). Basic and ultrabasic rocks contain more magnetite than less basic rocks. Sedimentary rocks contain appreciably less magnetite than the basic rocks (Breiner, 1973).

Since magnetite is so resistant to weathering, it appears in soils. The magnetic properties of many soils appear to have a strong relationship to the relative amounts of contained magnetite. In addition, organic action, particularly in high humus soils, is thought to be responsible for the formation of maghemite from nonmagnetic forms of iron oxide (Breiner, 1973). Note, also, that humic acids are good reducing agents. This action could account for the co-presence of magnetic and sedimentary uranium minerals.

The magnetic susceptibility is given in older literature in cgs units (rather than SI or MKS units):

$$B = \mu_0 \left(1 + 4\pi\chi_{cgs}\right) H \tag{4.9}$$

where χ_{cgs} = the magnetic susceptibility in cgs units.

Thus,

$$\chi_{SI} = 4\pi\chi_{cgs} \tag{4.10}$$

In either case, the magnetic susceptibility is dimensionless.

4.6.2 Magnetic Susceptibility of Geologic Materials

The magnetic susceptibility of magnetite is 0.3 cgs (3.8 SI units). It can vary from 0.1 to 1.0 cgs (1.3 to 13 SI units), depending upon grain size and other properties (Breiner, 1973). Thus, a 1% concentration, by volume, of magnetite would give a susceptibility, χ_{cgs} of 3×10^{-3} cgs. This, of course, could be widely variable. Figure 4.6 is from McNeill et al., and shows the relationship between magnetite content and magnetic susceptibility for various rocks and samples. Table 4.3 shows typical magnetic susceptibilities. Figure 4.7 shows the magnetic susceptibility of soil samples from three areas (from McNeill et al., 1996 and Cook and Carts, 1962). Note the wide variations of the susceptibility values. Cook and Carts found no correlation between soil color and magnetite content, unless the magnetism was extremely high. Thus, it appears that some rocks may be correlated with magnetic susceptibility as the correlation of soils is too poor to be used for identity.

Many soil samples exhibit values of χ_{cgs} in the range of 10^{-5} to 4×10^{-5} cgs units (1.20×10^{-4} to 5×10^{-4} SI units). When using a simple two-coil sonde (a transmitter and a receiver coil separated by a center-to-center distance of L_s (Figure 4.8) and these low magnetic susceptibility values, temperature compensation is needed.

Further, when the sample has a high electrical conductivity (low resistivity), such as one finds in most sediments, the out-of-phase, or conductivity, component dominates the signal. This demands that the signal be corrected for this component in these situations. Thus, conductivity must also be measured.

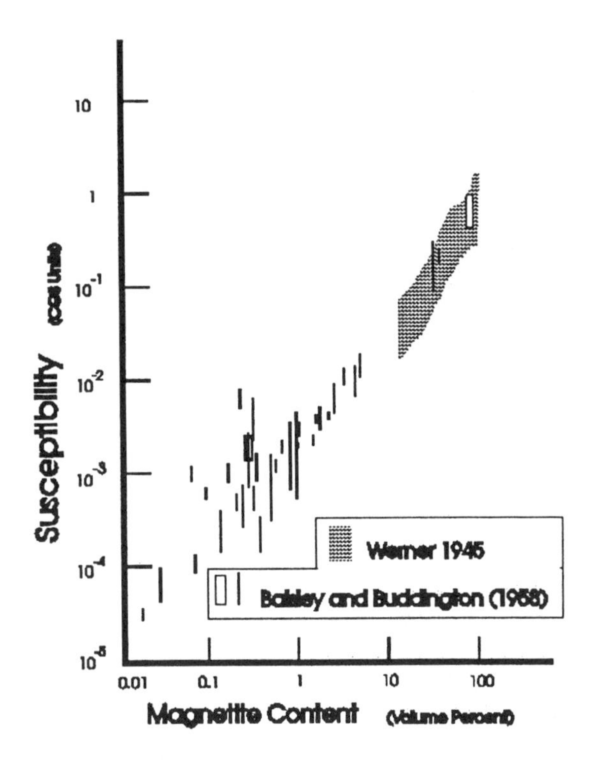

FIGURE 4.6
The relationship between magnetite content and magnetic susceptibility. (Courtesy of The Environmental and Engineering Geophysical Society.)

TABLE 4.3
Typical Magnetic Susceptibilities

Rock Type	Range (cgs units)
Altered Ultrabasic	10^{-4} to 10^{-2}
Basalt	10^{-4} to 10^{-3}
Gabbro	10^{-4}
Granite	10^{-5} to 10^{-3}
Andesite	10^{-4}
Rhyolite	10^{-5} to 10^{-4}
Shale	10^{-5} to 10^{-4}
Schist and other metamorphic rocks	10^{-4} to 10^{-3}
Most sedimentary	10^{-6} to 10^{-5}
Limestone and chert	10^{-6}

From Breiner, 1973.

4.6.3 Field Tests

McNeill used a two-coil downhole array to run tests of the feasibility of magnetic susceptibility logging. One coil (Tx in Figure 4.8) was the transmitter. It was energized with a current of 39.2 Hz. Alternating currents were used to

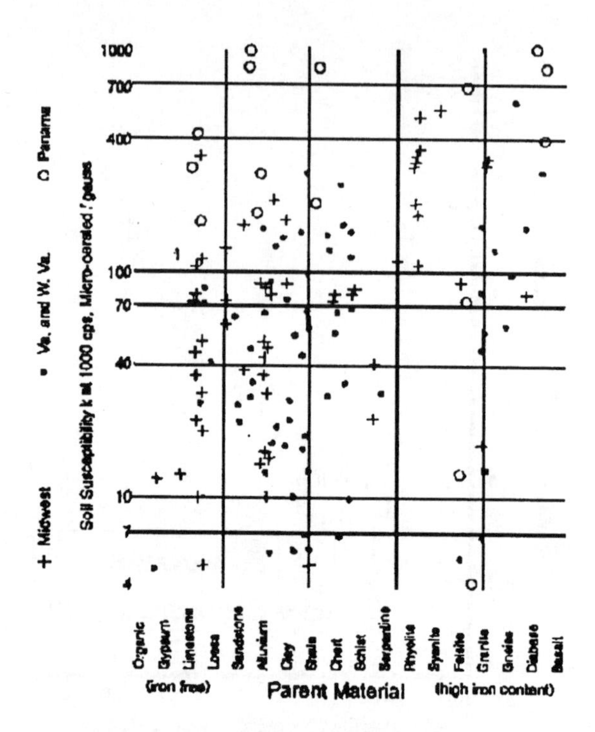

FIGURE 4.7
The relationship between soil samples from three areas and their magnetite content. (Courtesy of The Environmental and Engineering Geophysical Society.)

avoid the effects of residual natural magnetic fields in the earth. The magnetic field around a coil with an iron core is shown in Figure 4.9. The resulting magnetic field was sensed by the receiver coil, Rx. The intercoil spacing, l, was 50 cm (19.7 in.).

In the absence of any magnetically active material, the flux density, B, at the receiver is

$$B_0 = \frac{\mu_0 M}{2\pi L_s^3} \tag{4.11}$$

where

M = the transmitter dipole moment, the current, I, times the area, m^2
L_s = the inter-coil spacing, center-to-center, in meters

When the ground has a finite magnetic susceptibility and conductivity, the measured magnetic field is a complicated function of the ground magnetic susceptibility, conductivity, current frequency, and coil spacing. If the conductivity is low, as in most igneous and metamorphic rocks, the signal will contain a small in-phase component from it. If the conductivity is high and the magnetic susceptibility is low, as in many sediments, the electrical conductivity must be measured. Further, since the flux density at the receiver is

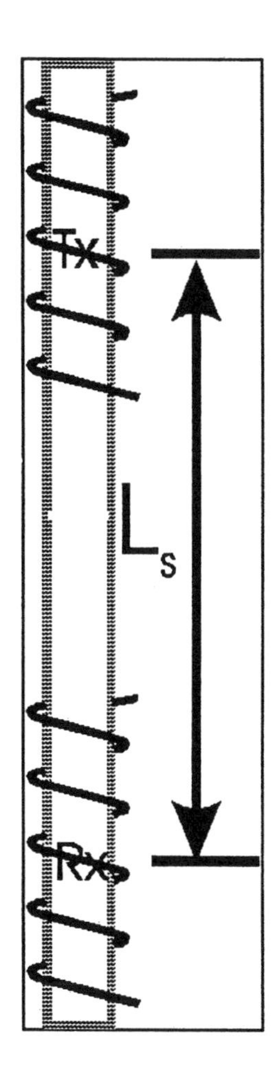

FIGURE 4.8
The coil arrangement of a magnetic susceptibility tool.

an inverse function of the cube of the spacing, temperature changes must be taken into account.

In magnetic material, at low conductivity, the in-phase component of the resulting signal is

$$B_0 = \frac{\mu_0(1+\chi_{SI})M}{2\pi L_s^3} \tag{4.12}$$

Therefore:

$$\chi_{SI} = \frac{2\pi L_s^3 B}{\mu_0 M} - 1 = \frac{B}{B_0} - 1 \tag{4.13}$$

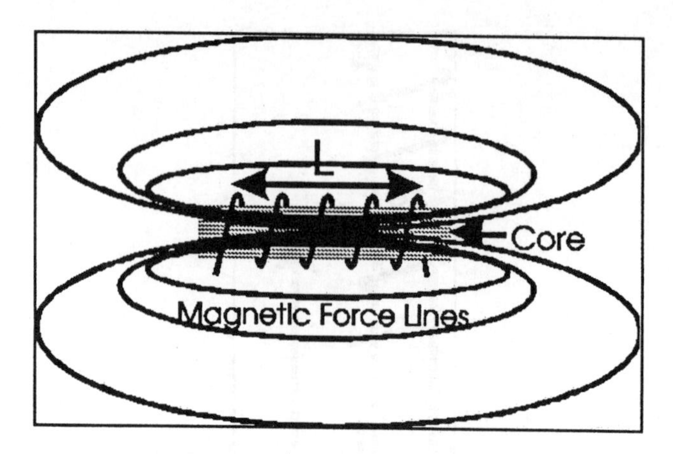

FIGURE 4.9
The magnetic field around an iron-cored coil.

Thus, if B_0 is accurately known, this provides a solution for an inductive magnetic susceptibility meter. B_0 can be obtained by calibrating the sonde high enough above the ground (~2 m) and away from any vehicles.

The U.S. Department of Energy (DOE) ran field tests of a two-coil magnetic susceptibility system in the Powder River Basin, in Wyoming (the Morton Ranch) around 1970. This author does not have notes, literature, nor examples from this test, but describes them from memory — we were caught in a blizzard on the ranch.

The DOE tests were in shallow sediments (200 to 300 ft, 60 to 90 m) in known uraniferous zones. Both were viable (containing ore-grade uranium geochemical cells) and relict, oxidized zones. The presumption was that the viable cells would have trace amounts of co-deposited magnetic iron compounds (i.e., magnetite) and that these compounds, in the oxidized zones, would be hematite-like. Therefore, the viable zones were expected to have effectively higher levels of magnetic susceptibility. They bore out the suppositions very well. Temperature drift problems were, however, quite severe.

McNeill et al. (1996) reported their tests in hard rock Precambrian zones that were overlain with Paleozoic sediments. These tests tend to confirm the experiences of the DOE and are reported in detail for three areas. Figures 4.10, 4.11, 4.12, and 4.13 show their results in one field at Bells Corner, Ontario.

Figure 4.10 shows the complete log and the corresponding geological column. It shows some temperature drift, in spite of the built-in temperature compensation. Referring to Equation 4.12, note that the intercoil spacing, L, appears as a power of three in the denominator. This would make this type of device extremely sensitive to temperature changes. McNeill et al. ascribe the excursions at 22, 33, 36, and 38 m as probably being caused by residual metal in the hole. They correctly state that the shape of each anomaly (an inverted peak) indicated that the artifact was smaller than the intercoil

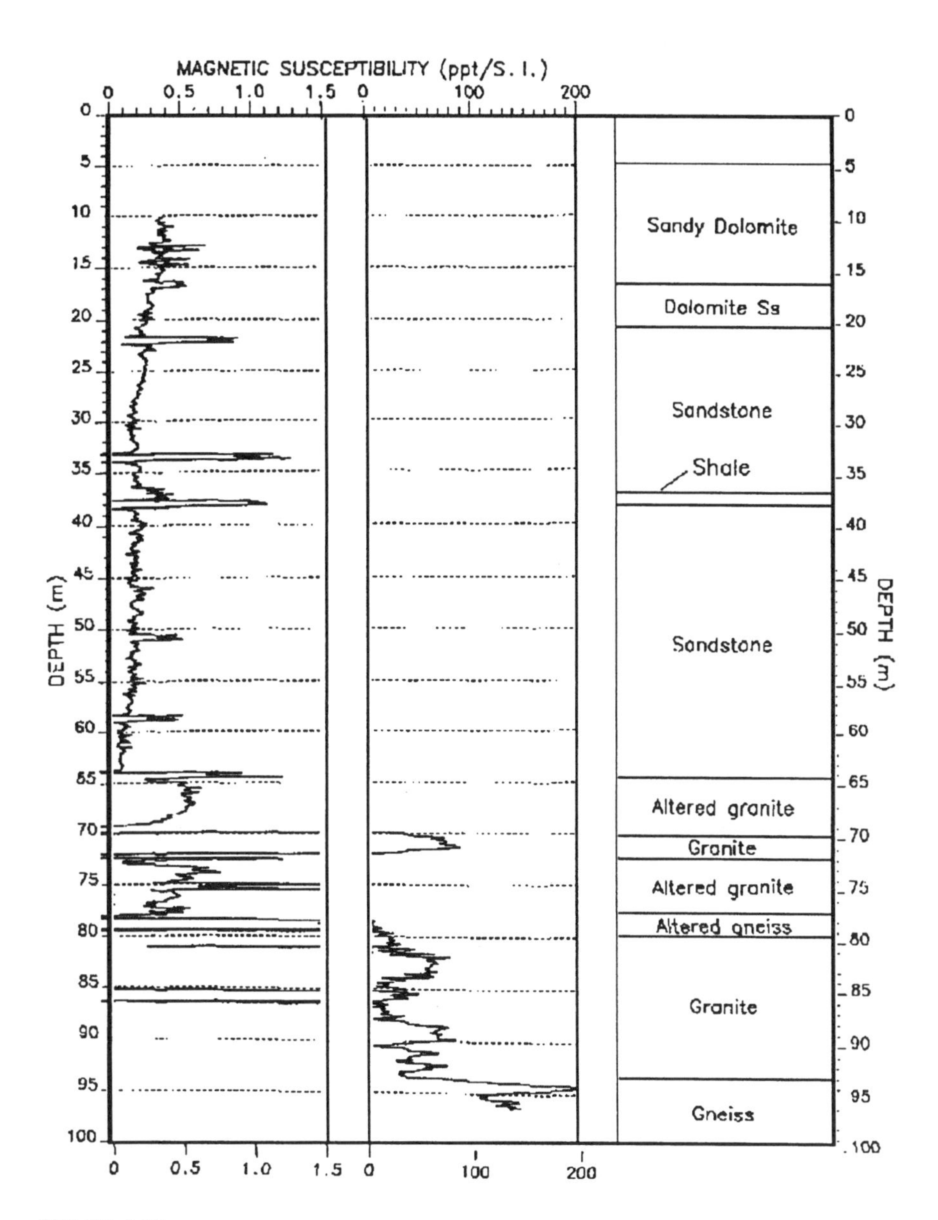

FIGURE 4.10
A log of magnetic susceptibility at Bell's Corner, Ontario. (Courtesy of the Environmental and Engineering Geophysical Society.)

spacing. This effect is similar to that observed with the two-electrode normal resistivity system.

Figure 4.11 is a view of a portion of sediments in another borehole, GSC-FD94-3. The geologic column is also shown. Figure 4.12 is another view of sediments in borehole GSC-FD94-4.

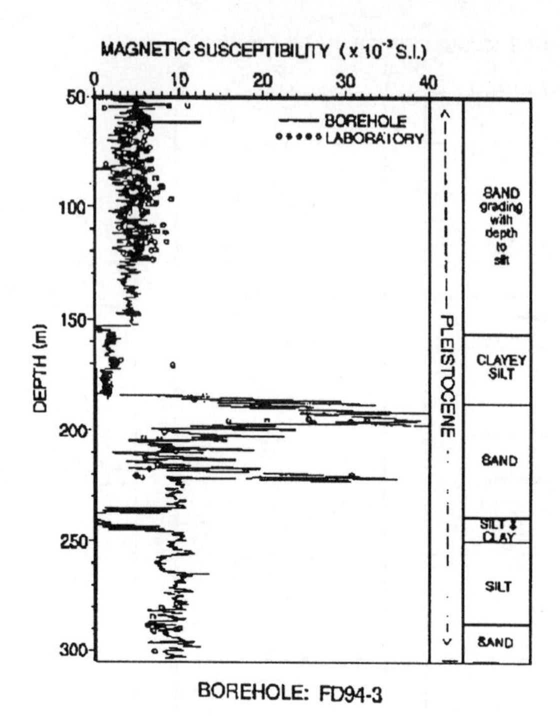

FIGURE 4.11
A magnetic susceptibility log in sediments. (Courtesy of The Environmental and Engineering Geophysical Society.)

This zone is the boundary layer at the interface between the Paleozoic overlying sediments and the Precambrian basement granites.

If the temperature drift is taken into account, the magnetic susceptibility values of the log agree well with those shown for sedimentary and granitic materials in Figure 4.13. Note on the logs, the wide difference in the values for granitic materials and for sandstone, of a factor of 350 or more. This type of contrast can make this technique more valuable than gamma ray, resistivity, and density methods for lithological and formation identification in metamorphic and mixed zones.

4.6.4 Laboratory Methods

The measurement of the magnetic susceptibility of samples in the laboratory is an old practice. It is a method taught in high school physics classes. The methods and calculations are identical to those given earlier in this chapter for downhole work.

Laboratory methods frequently use a single toroid and measure the self-inductance, Λ, of the coil and its contents:

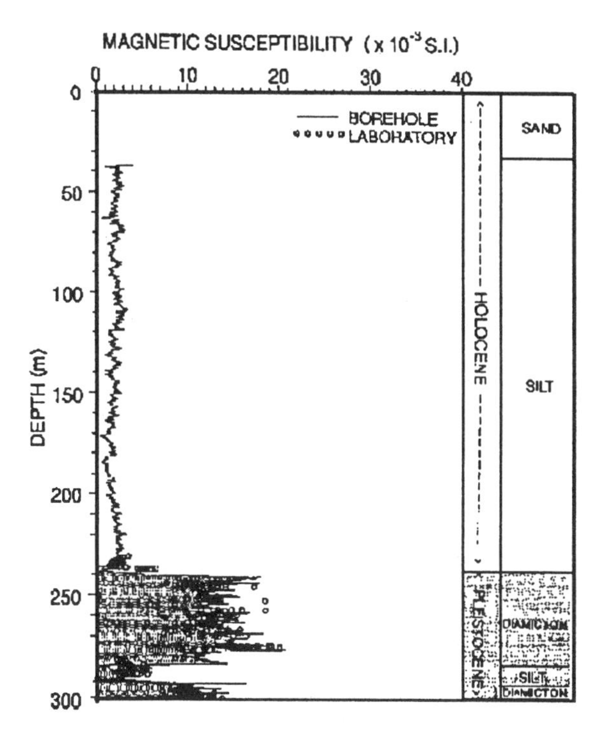

FIGURE 4.12
A log in the sediments of hole GSC-FD94-4. (Courtesy of The Environmental and Engineering Geophysical Society.)

$$\Lambda_m = \chi_m \mu_0 \frac{AN^2}{L} = \Lambda_0 \chi_m \tag{4.14}$$

where χ_m = the magnetic permeability of the toroid, and its contents and is a factor representing the change of the self-inductance when a sample is placed in the toroid:

$$K_M = \frac{\mu}{\mu_0} = 1 + \frac{\chi}{\mu_0} \tag{4.15}$$

Λ_M = the self-inductance of the toroid in a vacuum and removed from any other material.

As in the downhole systems, an alternating current is used to activate the coil in order to cancel out the effects of residual or permanent magnetism in the sample.

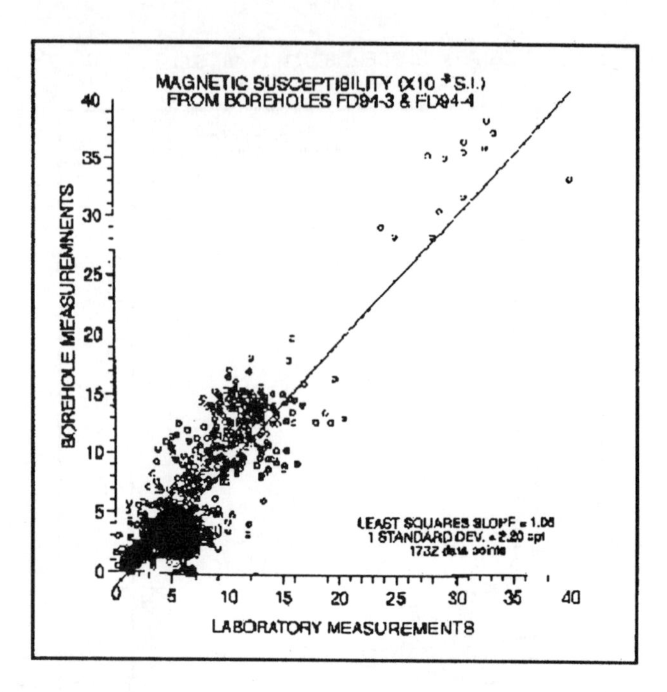

FIGURE 4.13
The agreement between borehole and laboratory measurements. (Courtesy of The Environmental and Engineering Geophysical Society.)

4.7 Radar

Pulse radar methods, using high frequency electromagnetic pulses, have found some application in nonhydrocarbon work. These methods were first proposed in 1927 by Hulsenbeck et al. They are used primarily in low conductivity formation (i.e., crystalline rocks) to locate discontinuities. The discontinuities may be high conductivity, such as conductive mineral bodies, or low conductivity, such as gas-filled fissures or mine drifts. These methods can be sensitive enough to detect rock types.

One form of geophysical, borehole radar uses a dipole transmitting antenna in one borehole and a receiving antenna in another some distance away. This distance may be from 10 to 200 m, as shown in Figure 4.14 and is the hole-to-hole or cross-hole method. The recording is made as distance (borehole depth) against time. This is the same type of presentation as used in seismic work. The dispersion rate can be quite low in a low conductivity crystalline rock, such as quartzite. Pulse repetition rates are high (~50 kHz). Pulse frequencies are from 50 to 500 MHz. The pulse to the transmitter triggers the receiver and the clock via a fiberoptic connecting cable.

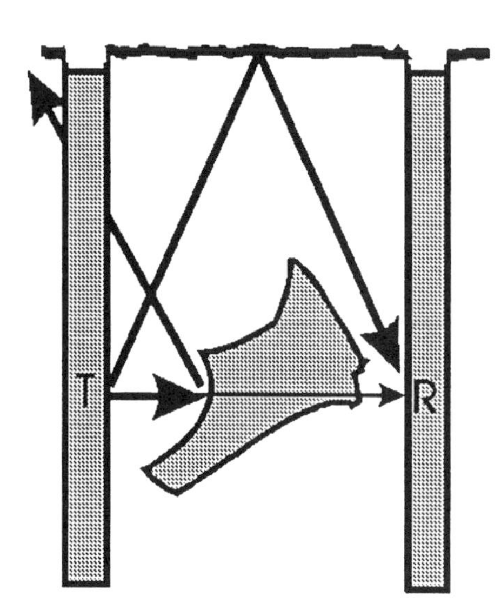

FIGURE 4.14
The cross-hole radar method.

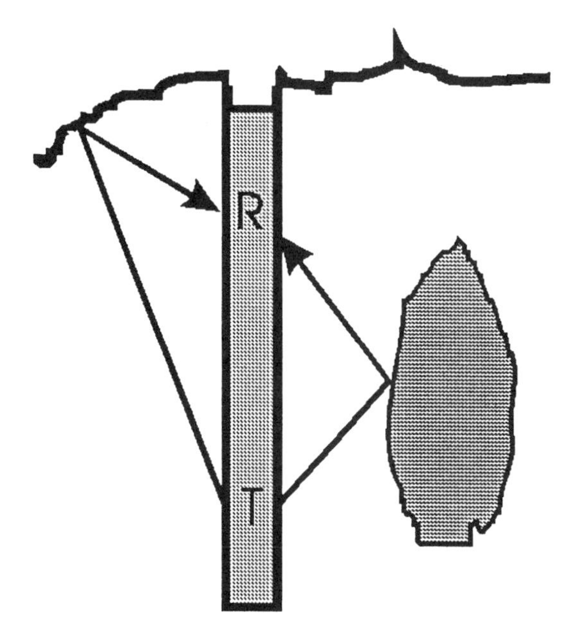

FIGURE 4.15
The reflection radar method.

A second method is shown as Figure 4.15 and uses much the same equipment. It has both the transmitting and the receiving antennae in the same hole, however. This method has the advantage that the signal reflected from the target is stronger than the one transmitted through a conductive body. Signals reflected from the surface also are less problematic.

The transmission velocity of the electromagnetic pulse from the transmitting antenna, in a homogeneous medium, is a function of the medium permittivity. The amplitude of the received pulse is a function of the average conductivity of the medium through which it passes. A pulse passing through an intervening conductive body will be reduced in amplitude by reflection from the surface of the body and by absorption within the body.

Investigations have been conducted using the antennae on the surface of the ground. The penetration of the signals is shallow, however, but images can be obtained. Ground penetrating radar (GPR) has been used to locate pipes, measure overburden thickness, and investigate construction sites. The University of Waterloo (Toronto) and the U.S. Environmental Protection Agency have used the method to trace the movement of immiscible fluids near the surface.

The high frequency of the radar pulse allows a good resolution. The resolution can be better than a wavelength because the time detection method can resolve details within the pulse. With wavelengths of 2 to 5 m, resolutions with one type of equipment are reported to be better than 0.5 m (Sandberg et al., 1989).

5

Mechanical Wave, Acoustic Methods

5.1 Introduction

The acoustic methods were discussed in *Standard Methods of Geophysical Formation Evaluation*. The reader is referred to that chapter for background information about the acoustic phenomena.

Acoustic (or mechanical wave) methods have been a major tool for petroleum exploration and development. These methods are being used increasingly in nonhydrocarbon work. A reason for this development has been the application of acoustic methods to hard rock environments and the application of tomographic techniques. Some of the methods discussed in this chapter were developed specifically for and used primarily in the oil industry. They appear, however, to have some important applications in nonhydrocarbon work. Therefore, they have been included in this volume.

All acoustic methods introduce a mechanical disturbance into the formation material. The disturbance is usually an energy pulse, although continuous wave devices have been investigated and some are in use. The pulse may originate on the surface or in a borehole and may originate from any one of several types of sources: piezoelectric transducers, magnetostrictive devices, mechanical thumpers, hand-held hammer and pad, explosions, or electrical discharges. Continuous wave systems usually use a mechanical vibrator, such as an unbalanced wheel. The physics are similar to that of physical optics. Acoustic methods are used downhole, on the surface, a combination of the two, between two holes, and in the water. The names of the various devices are numerous. However, sonic, acoustic, seismic, and (often) mechanical all indicate acoustic methods.

5.2 Borehole Seismic Methods

Borehole seismic methods are used extensively in the petroleum business. They also have great potential use in nonhydrocarbon applications. There are, however, the usual drawbacks: the methods are not always well known, the existing equipment was designed for larger diameter oilwells, and the costs are high. If there is a demand, however, the services suitable for smaller lower cost, shallower nonhydrocarbon boreholes, will evolve.

The borehole seismic methods are an adjunct or a necessary part of surface seismic techniques. They serve both as corrections and calibrations to the surface methods and as independent information sources. Their greatest value lies in their higher detail and more precise information — especially depth information. Compared to surface seismic methods, they cannot cover as wide a geographical area at as low a cost. The two most important aspects of borehole seismic methods are the "check shot" techniques and the "vertical seismic profiles" (VSP).

Surface seismic surveys and digital data processing and enhancement are used in the initial stages of exploration, especially in the petroleum industry. They could, as well, be used in such nonhydrocarbon exploration fields as coal, heavy minerals, hydrology (especially in limestone reservoirs), and engineering. This is especially true if the targets are fairly deep. That is, below the weathered surface layers. These targets are becoming more important as near-surface targets are used up, prove unsatisfactory, and/or environmental concerns increase. Deep targets then become attractive.

Even with digital processing there are severe limitations to the traditional surface seismic methods. These are twofold. First is the limitation imposed by the low frequencies that are used. These make details difficult to discern. Second is the lack of depth control. Seismic depth assignments depend partially upon assumed velocities for the acoustic waves. The petroleum industry has greatly overcome some of these obstacles by digital processing, creating synthetic seismograms from the downhole acoustic (porosity) and density logs, check shot techniques, and VSP.

Check shots and VSP consist of surface sound sources and downhole arrays of cable-mounted geophones in various configurations. In operation, repeated surface shots (noise pulses) and downhole readings are made at various depths throughout the borehole. Mechanical devices clamp the geophones to the wall of the borehole or to the casing. Readings may be made in either open holes or cased holes, as the casing shows little influence on the signal. Schlumberger Well Services, Inc. offers (for example) three different types of downhole arrays.

5.2.1 Vertical Seismic Profiling (VSP)

VSP is the downhole detection of surface-generated seismic sources. Because, in this method the energy wave travels only once through the surface weathered layers, the resolution of the VSP is better than that of surface methods. In addition, depth control is much more precise. This method can be used in open or cased holes (see Figure 5.1). Table 5.1 lists some of the advantages of the VSP method.

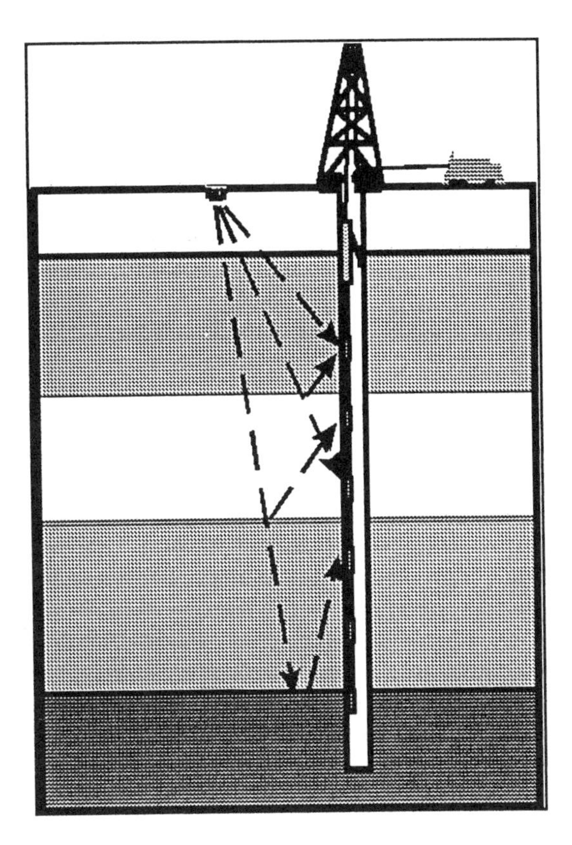

FIGURE 5.1
The principle of the VSP.

TABLE 5.1

Advantages of the VSP Method

1. VSP records a real seismic trace in the borehole, rather than relying on a synthetic seismogram
2. VSP displays the spectral content of the down-going seismic signal, as a function of depth
3. A precise link is established between the surface seismic results and the borehole logs
4. The higher frequency contents of the signals are recorded, rather than being absorbed by the second transit through the weathered surface layers
5. VSP has an improved resolution of subtle stratigraphic features around the borehole, compared to surface seismic methods
6. VSP records deep reflected signals which are not received at the surface. This allows better analysis of deep complex structures
7. VSP can provide a record of the band-limited reflection coefficient series

Courtesy of Schlumberger Well Services, Inc.

*5.2.1.1 Well Seismic Tools (WST)**

The WST uses four stacked geophone that are primarily sensitive to movement in the vertical direction (the pressure wave or p-wave).

5.2.1.2 Seismic Acquisition Tool (SAT)

The SAT tool has three mutually orthogonal geophones (which may be gimble-mounted for use in deviated wells) for three-dimensional operation. This arrangement provides an x, y, z system of references where each arriving ray can be represented by a vector. Among other applications, the ability to record and process signals in the three axes allows recording and interpretation of shear waves, salt proximity surveys, and long-offset VSP surveys.

5.2.1.3 Downhole Seismic Array Tool (DSA)

The DSA tool uses eight sensor packages (shuttles) that are positioned along an insulating, multiconductor bridle cable at intervals of up to 50 ft (15 m). The sensor package contains a vertical geophone for signal acquisition, a magnetic clamping device to secure the package to the casing, a shaker element to generate mechanical vibrations for reference, and electronic circuitry to transmit the signals to the cartridge. In the cartridge, the signals pass through anti-liaising (anti-cross-coupling) filters, sample-hold circuits, and multiplexers. They are then digitized and telemetered to the surface.

5.2.1.4 Sources

Any seismic source may be used for VSP surveys. Vibroseis is a popular source. Schlumberger, for example, uses both air guns and Vibroseis. Offshore work frequently involves air-guns because of safety, economy, reliability, broad frequency spectrum, simple signature, and ease of transport. Airguns can be fired in array for more intense sound pulses. Further, the wave shape may be modified and tailored by means of the source design.

5.2.1.5 Technique

VSP makes use of an array of downhole geophones, usually spaced at short intervals (50 ft or 15 m) on a cable. Multiple surface shots (energy or noise bursts) are made each with the array at a different depth in the borehole. A surface geophone or hydrophone is often used to time the initiation of the pulse. Transit time is then measured from the first of the burst at the surface detector to the arrival at each of the downhole detectors. Several shots may be made at the same level and the results stacked to improve the quality of the signal.

---

* Sections 5.2.1.1, 5.2.1.2, and 5.2.1.3 are courtesy of Schlumberger Well Services, Inc., taken directly from *Log Interpretation Principles/Applications*, 1989; a Schlumberger sales document.

If the downhole is deviated or if the source is offset substantially from the borehole head, the transit time must be corrected to the true vertical depth. Oilfield VSP measurements are usually corrected to the surface seismic reference datum, if the VSP source is above or below the seismic datum level.

5.2.1.6 Check Shots

Check shot measurements are used to correct the velocities obtained from the acoustic borehole log transit time values. The adjusted acoustic log is then used to translate the surface seismic times into depths. It is also used to calculate acoustic impedances in synthetic seismograms.

The seismic velocities may be different from those obtained by integration of the acoustic log transit times for several reasons:

1. Differences between seismic frequencies (~50 Hz) and acoustic log frequencies (~20 kHz); when acoustic log transit times are used to calculate velocities for the synthetic seismograms may result in substantial differences.
2. Borehole effects, such as formation alteration (due to drilling), may decrease acoustic velocities.
3. The acoustic log measurement is fundamentally different from the seismic measurement in both signal character and path.

Thus, seismic times are normally referenced from check shots and the acoustic logs are adjusted accordingly.

The VSP records a surface shot that has only passed once through the low velocity, weathered surface. It also records the down- and up-going signals simultaneously, allowing analysis of the formation changes with depth.

The total wave train at each downhole detector consists of:

1. Signals arriving from above the detector (down-going). These are both the direct arrivals and the down-going multiples (reflections from the surface).
2. Signals arriving from below are the up-going signals. They consist of direct reflections and up-going multiples.

5.3 Borehole Televiewer

The borehole televiewer patents are held by Mobil Oil Company. It is an acoustic device that responds to the irregularities of the borehole wall. Its name is rather misleading, as it is not a visual (television) device, but rather, uses mechanical pulses (more akin to the sonar family). Use is not limited to boreholes filled with transparent fluids. The device can be used in water, water-based muds, emulsions, and oil-based muds. The pulse from the transducer may travel "several feet" into the formation from the borehole. The signal involves nearly 10^7 data points per foot of hole (3.2×10^7 data points per meter of borehole).

The borehole televiewer uses a rotating transducer which emits a train of high frequency mechanical pulses. These pulses reflect from the formation material around the borehole wall. The travel time of the first arrival, back at the sonde, is detected. Thus, the borehole diameter can be determined. The signal may be dispersed by any irregularity of the borehole wall as well as by fractures, discontinuities, and fissures. The scan can also form a picture of the texture of the borehole wall from the resulting amplitude of the return signal. The casing is relatively transparent to the acoustic signal. Figure 5.2 shows the principle of the acoustic borehole televiewer.

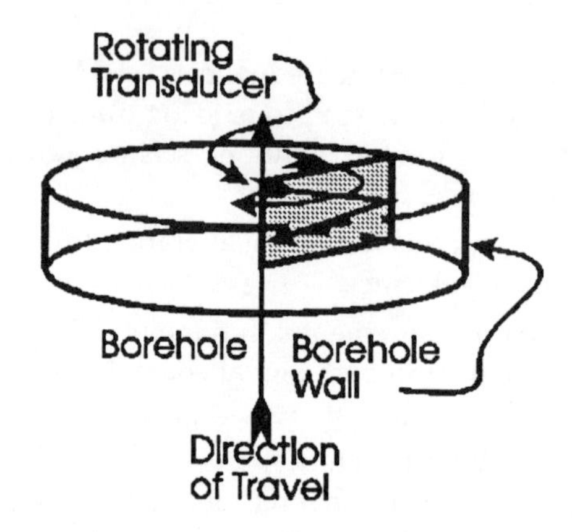

FIGURE 5.2
The principle of the borehole televiewer. (Adapted from Amoco.)

Presentation is usually a two-dimensional, continuous strip, as a function of depth. The horizontal dimension is the radial position around the borehole and is usually presented as north-east-south-west-north (N-E-S-W-N), left to right.

Results from the televiewer are poor in elongated, collapsing, and rugose boreholes. This environment, of course, is common in fractured zones. Recent models have a much better performance than earlier ones.

A modification of the original televiewer, by Amoco Production Research, Inc. is identified by the name Volumetric Scanning (VS). This device records the travel time and also the signal amplitude at the receiving transducer. The amplitude is recorded on a gray scale as a wall character trace. The travel time is also recorded (caliper). Figure 5.3 shows the principle of the VS system.

These devices are large and not well suited to the usual mineral-type borehole. They are worth considering, however, in some engineering and commercial water-well holes. Figures 5.4, 5.5, and 5.6 show some of the results of

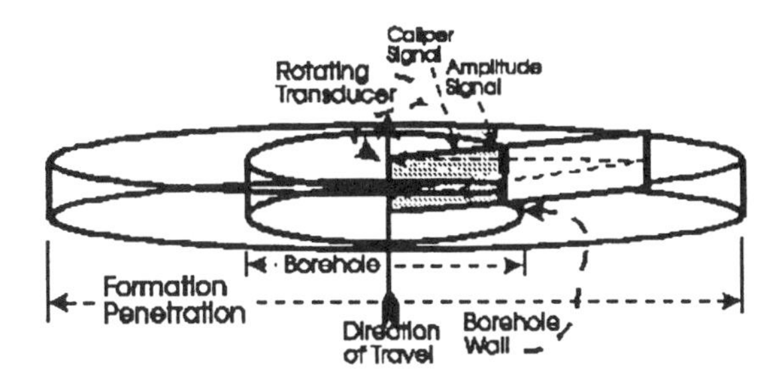

FIGURE 5.3
The principle of the volumetric scanning system. (Adapted from Amoco.)

the borehole televiewer system. Figures 5.4 and 5.5 are the same section, but Figure 5.5 has a deeper penetration than Figure 5.4. Figure 5.6 shows the same section in four panels, each rotated by 90 degrees from the previous one.

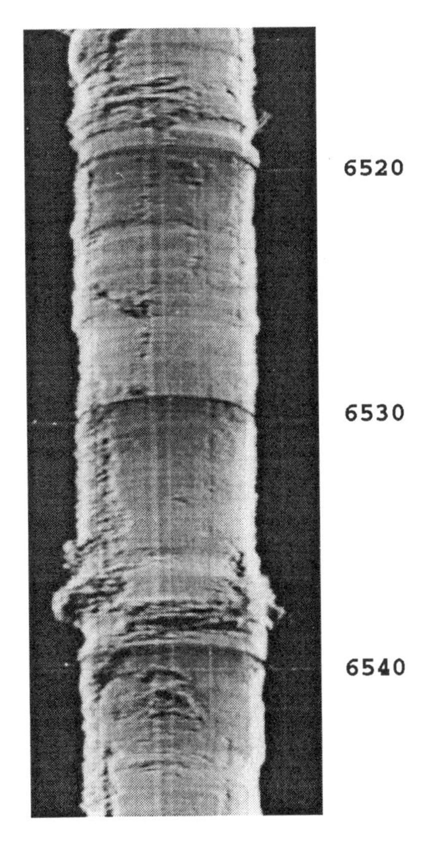

FIGURE 5.4
A televiewer view of a well section.

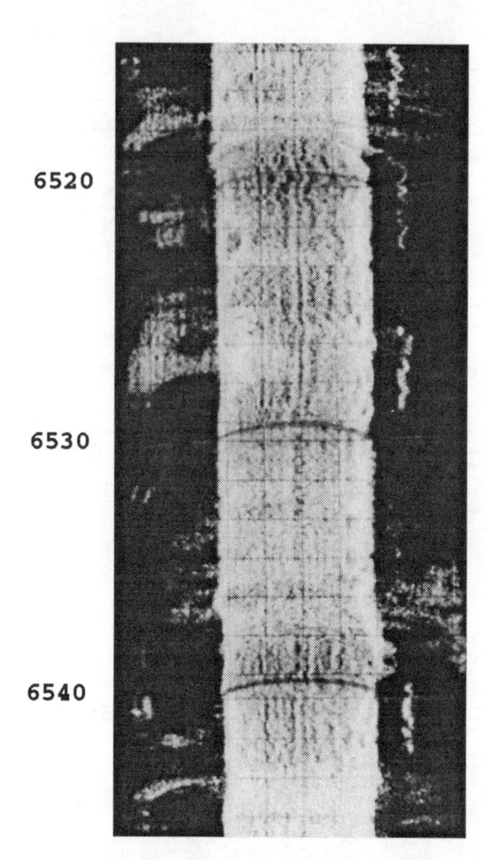

FIGURE 5.5
The same section as in Figure 5.4, but with deeper penetration.

5.4 Array Sonic Systems (AIT)*

The Array Sonic System was discussed in detail in *Standard Methods of Geophysical Formation Evaluation*. This device should be seriously considered when rock mechanic problems are being investigated. The waveform presentation, from the Schlumberger Well Services manual, *Log Interpretation Principles/Applications, 1989,* is presented here (it also appears in *Standard Methods of Geophysical Formation Evaluation*) to emphasize its clear identification of the P-wave, S-wave, and the Stoneley wave with this type device (see Figure 5.7). This allows a better rock mechanic analysis than do some of the older presentations and devices. This type of analysis is especially valuable in sedimentary environments. It should prove to be as valuable in hard rock environments, but, as yet, the documentation on that aspect does not exist.

* Trademark of Schlumberger Well Services, Inc.

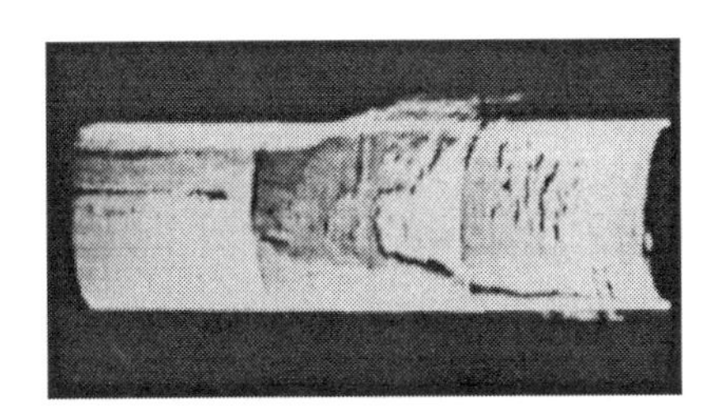

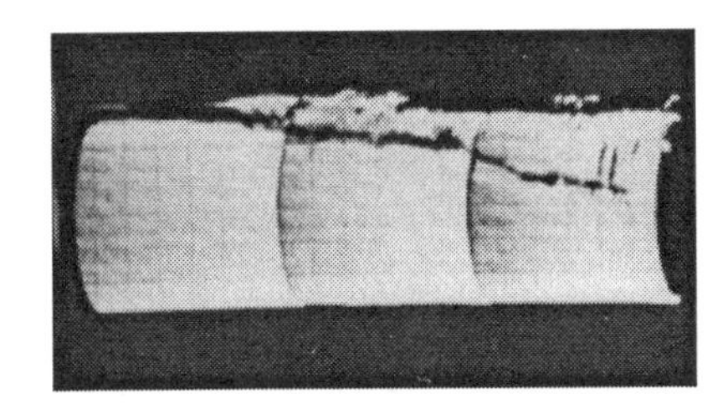

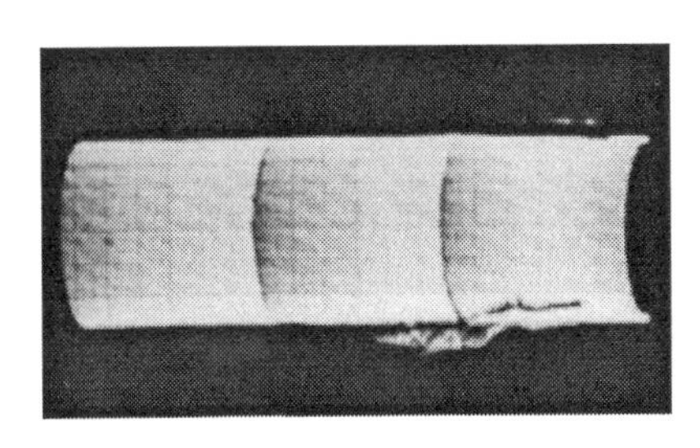

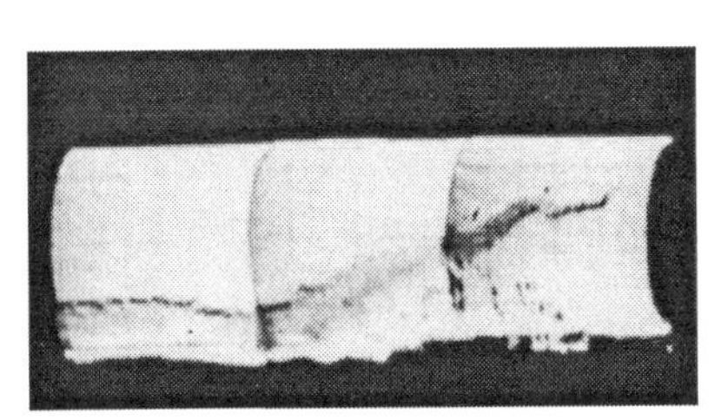

FIGURE 5.6
The televiewer image rotated to show different sides. (After Amoco.)

Consider combining this device with information obtained by the following systems: surface seismic, a borehole caliper, a resistivity dipmeter (may be of limited use in a hard rock environment), a tool orientation device, a borehole televiewer, a borehole television, the microscanner log, the VSP, the formation density log, the natural gamma ray and/or spectrograph, the temperature log, a gravity survey, surface magnetic field survey, magnetic susceptibility log, and a sibilation (noise) log.

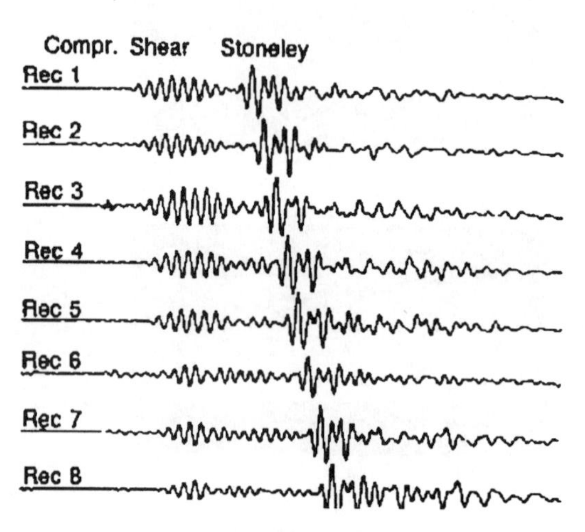

FIGURE 5.7
Array sonic waveforms.

The above logs and their various combinations allow good analysis of rock type, rock strength, the presence of fractures or caves, and faulting. These features are vital to engineering, construction, and mining. Rock mechanics will be discussed in a later chapter, after the various measurements have been discussed.

5.5 Long-Spacing Acoustic Systems (LSS)*

The long-spacing acoustic systems are represented by Schlumberger's Long Spacing Sonic Tool (LLS). These have two transmitters and two receivers, with spacings of 8 ft (2.44 m) and 10 ft (3.05 m) or 10 ft (3.05 m) and 12 ft (3.66 m). The LSS tool measures the transit time at a greater depth laterally into the formation than does the usual or standard acoustic porosity system. The deeper signals are more likely to give readings free from alteration (due to invasion), relaxation (due to drilling damage), and enlarged and overgauged holes than the shallow, standard porosity system. These measure-

* Trademark of Schlumberger Well Services, Inc.

ments are especially valuable when acoustic travel time readings are to be used in conjunction with seismic data.

The LSS tool uses a computer-derived borehole compensation method to save excessive downhole tool length. It uses two transmitters (T1 and T2) and two receivers (R1 and R2). Figure 5.8 shows the principle of the LSS system. A first reading (T1 to R1 minus T1 to R2 = t_1) is saved. A second reading (T1 to R2 minus T2 to R2 = t_2) is taken 9.67 ft (2.95 m) later. These are combined:

$$t = \frac{t_1 + t_2}{2s} \tag{5.1}$$

where s = the distance from R1 to R2 (2 ft or 0.61 m).

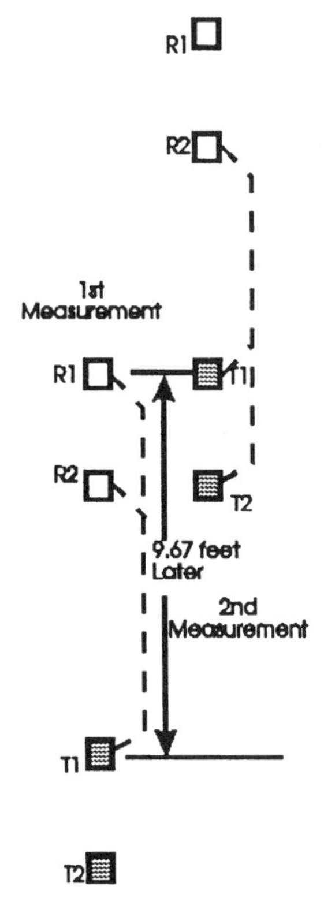

FIGURE 5.8
The principle of the LSS System. (Courtesy of Schlumberger Well Services, Inc.)

This LSS system may provide better acoustic values in permeable zones than the standard acoustic porosity systems. In nonpermeable zones, however, there is little advantage to the LSS system over the BHC systems. Further, the generally lower quality of the typical mineral-project drilling mud may result in invasion of permeable sediments deeper than even the LSS

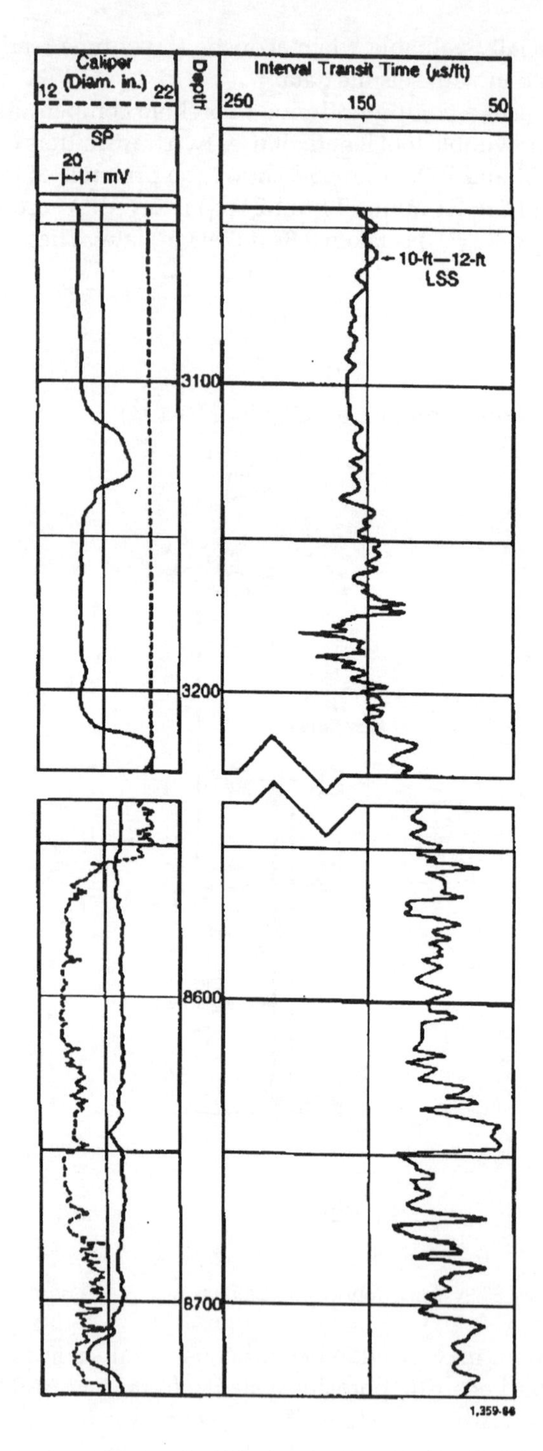

FIGURE 5.9
A log from the LSS System. (Courtesy of Schlumberger Well Services, Inc.)

measurement reaches. Also, the existing systems are mostly designed for large diameter oil wells. Most of these tools require a 6 in. (15 cm) or larger diameter borehole. Figure 5.9 shows a log from the LSS system.

If the sonde is tilted, with respect to the axis of the borehole, the same amount for both readings, the reading is equivalent to the borehole compensation of the BHC systems. It should be remembered, too, that the BHC feature *only* compensates for the axial tilt of the sonde in the borehole. *It does not compensate for enlarged or rugose portions of any borehole.*

6

Thermal Methods

6.1 Introduction

Downhole temperature measurements are potentially quite valuable, because the thermal environment and other physical environments are closely related. The thermal environment of the formation material near the borehole has been disturbed to such a degree, however, that long periods of time are generally required for it to stabilize after drilling. The temperature determinations generally made in a borehole are merely the thermal state of the borehole fluid at the time of logging. This is seldom at the true formation temperature and is probably changing appreciably with time. While the measured borehole temperature is that of the borehole fluid, it is still influenced greatly by the formation thermal environment. These temperature measurements and determinations, however, are useful and should not be discounted. Additionally, with a little effort and care, some valuable quantitative thermal information about the formation can be determined. Table 6.1 lists some of the uses of borehole temperature measurements.

TABLE 6.1

A Few Uses of Temperature Measurements

Temperature measurements are used for
1. Making corrections to solution resistivities
2. Determination or estimation of formation temperature
3. Gas entry location
4. Liquid entry location
5. Cement top location
6. Detection of fluid loss
7. Detection of fluid communication
8. Correlation with the electrical resistivities

6.2 Principles

Because the interior of the earth is hot, compared to the exterior surface, the thermal potential field causes a flow of heat from the interior to the exterior and outward to interstellar space. Some thermal radiation is received from space, primarily from the sun. Local surface variations are well known. Severe local subsurface variations are usually evident near active and dormant volcanoes and other tectonic disturbances. There are measurable, quasi-predictable temperature gradients through the upper layers of the earth where exploration, drilling, and production takes place.

The heat flow at any point in the earth material and the temperature gradient through any length of that material depends upon the temperature difference across the material, the area of the cross section of the sample, the thermal conductivity of the material, and the rate of flow of the heat energy. These things are analogous to electrical parameters: temperature, temperature field, and temperature gradient to voltage, voltage field and voltage gradient, heat flow to electrical current, and thermal conductivity to electrical conductivity.

Heat flow is described by Fourier's law:

$$\frac{dQ}{dt} = -K_T A \frac{dT}{dx} \tag{6.1}$$

where

- dQ/dt = the quantity of heat flow per unit time.
- A = the area of the cross section of the material through which the heat is flowing.
- K_T = the conversion constant (the thermal conductivity of the material).
- dT/dx = the temperature gradient (with respect to path length) along the path of heat flow (the thermal field).

Notice the close relation of Equation 6.1 to that of the flow of electrical current through a conductive medium:

$$\frac{dQ}{dt} = -\sigma A \frac{dV}{dx} \tag{6.2}$$

where

- dQ/dt = the quantity of electrical current per unit time.
- A = the cross-sectional area of the sample.
- σ = the conversion constant (electrical conductivity of the material).

dV/dx = the voltage gradient or field along the direction of flow of the current.

Fluid flow is similarly related. Darcy's law of fluid flow is:

$$\frac{dQ}{dt} = K_p A \frac{1}{\mu} \frac{dP}{dL} \tag{6.3}$$

where

dQ/dt = the quantity of fluid flow per unit time.
A = the cross-sectional area of the sample.
K_p = the conversion constant (permeability).
μ = the fluid viscosity.
dP/dL = the pressure gradient.

6.3 Borehole Temperature

The drilling of the borehole inevitably results in severe thermal disturbances to the formation volume immediately adjacent to the hole. The disturbance reverses as soon as circulation is stopped and the borehole temperature begins to change. After a period of days, weeks, or months, the volume returns to its original thermal state. The rate of recovery is most rapid during the initial time, in the first few hours after circulation has ceased. This is, of course, because the temperature differences are greatest at that time.

If enough time is allowed to elapse between the formation temperature measurement and when circulation is stopped or casing is set, satisfactory formation temperature measurements can be made and the temperature gradient and thermal conductivities calculated. The time needed for the thermal environment to stabilize around a well, however, can be days, weeks, or even months.

Because the thermal environment is so disturbed during drilling, casing, and production, the temperature measurements made during this time are normally used only for making temperature corrections to logged values. The temperature log is usually run after the circulation has stopped and/or after the well has been shut in for a stabilizing period of time ranging from a few hours to days or weeks. The stabilizing time period will vary for practical and economic reasons. The uses of the temperature log, however, change as the time period increases.

Borehole temperature logs are always run as the first run, with the sonde going down into the hole. This way, the system will measure an environment (mud column) undisturbed by the logging equipment. The temperature curve that is recorded, under these circumstances, is the temperature gradient of the borehole fluid. The longer the hole has remained undisturbed, the closer the gradient will be to the earth's thermal gradient. Small variations from the average gradient will be evident. These are due, in part, to local differences in the thermal conductivities of the formation materials. If the borehole fluid is gas (which is a common situation) and the hole is cased, the gradient will more quickly approach that of the earth's gradient, especially if the hole is filled with air or other gas during drilling.

The temperature at the formation level is approximately equal to the product of the mean temperature gradient, ΔT_m, and the depth, D, plus the starting temperature, T_s. If the formation temperature, T_f, is sought, the needed gradient will be the earth's temperature gradient at that location (not the borehole gradient). The starting temperature is the surface temperature, at that time. If the borehole temperature is desired (the usual case), the gradient is the borehole gradient and the starting temperature is the top-hole, borehole fluid temperature, T_{fl}. The latter is usually measured in the mud pit from the flow line jet. The "so-called" formation temperature, T_f, is

$$T_f = T_{fl} + \Delta T D \tag{6.4}$$

It is important to determine the borehole temperature as close to the time of running a log as possible.

Figure 6.1 shows a comparison of the temperature curve and the resistivity curve in the same Oklahoma well (Guyod, 1944). There is also a correlation between the washouts in the borehole and the temperature curve, provided that the temperature system has a short thermal time-constant and sufficient sensitivity.

6.3.1 Normal Uses

Formation fluid entry into the borehole can be detected because the entering fluid is seldom at the same temperature as the borehole fluid. Therefore, temperature gradient anomalies can frequently be used to detect fluid entry. The detection of gas entry is more positive than liquid entry because of the adiabatic cooling as the gas enters the lower pressure borehole. This is not frequent in nonpetroleum projects, but can take place when, for example, the entry of radon, helium, carbon dioxide, methane, and/or nitrogen occurs in uranium and coal projects. Motion of the fluid in the borehole and the direction of that motion can be determined by the temperature anomaly and its slewing, at the point of entry.

Portland cement (the material normally used for cementing casing in place) undergoes an exothermic reaction while setting. Cement and cement top location (in the formation/casing annulus) make use of this heat given off during the setting of the cement. Above the top of the cement the borehole temperature will show a sharp return to a near normal temperature gradient.

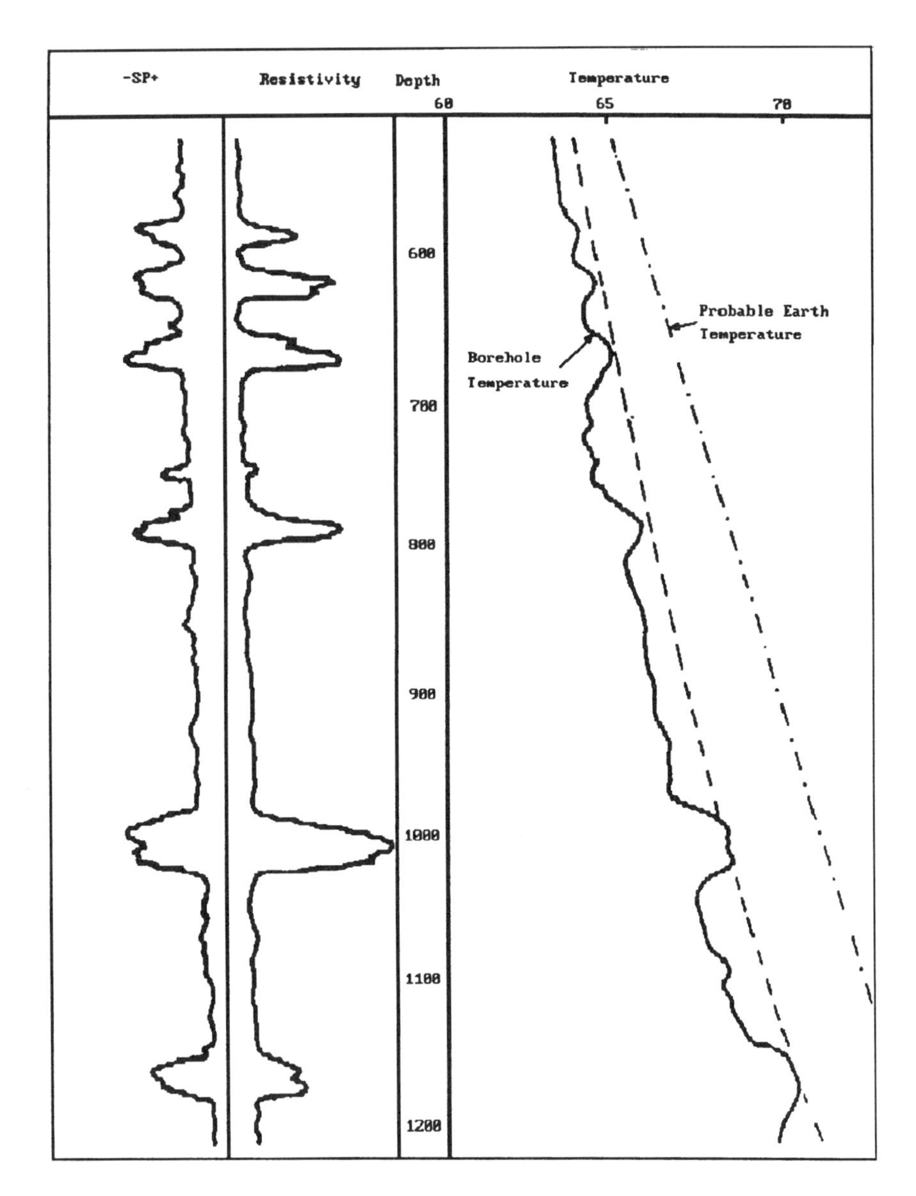

FIGURE 6.1
A comparison of the resistivity and temperature curves. (After Guyod.)

Detection of fluid loss and communication may involve the deliberate injection of liquid at a higher pressure than the formation and at a greatly different temperature. Then, temperature changes are monitored (see Figure 6.2).

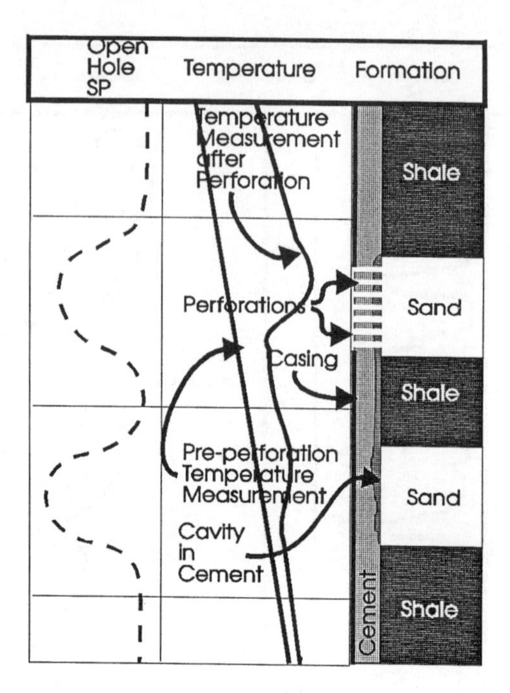

FIGURE 6.2
The effect of injecting warm water through perforations.

6.3.2 Instruments

In general, the downhole temperature instruments measure the borehole fluid temperature and temperature gradient. The simplest system uses a maximum reading mercury thermometer in the cable head of the logging system. The maximum temperature is assumed to be the bottom hole temperature (BHT). The flow line temperature is assumed to be the top temperature and is subtracted from the BHT to obtain the temperature change over the borehole. This is then divided by the total depth (TD) and assumed to be the average temperature gradient of the borehole.

With better methods, temperature is sensed with a thermocouple, a resistor bridge, or with thermistors. The voltage of a thermocouple is compared with that of an identical thermocouple that is kept at a constant, higher temperature. A thermocouple has the advantage of an inherently short thermal time-constant and a good predictable accuracy. The disadvantage is a low signal level and the fact that a reference temperature must be furnished. The use of these devices is mostly restricted to scientific work.

The resistor bridge is a device for comparing a temperature-sensitive resistor with a similar resistor that is not temperature sensitive. This is accomplished with a Wheatstone or other type of bridge circuit. This type of device is simple and easy to use. Signal levels can be quite good and signal-to-noise levels are satisfactory. In general, the temperature coefficient of resistance of the sensor and the references are the minor weak points of this type of device. Many of the materials used for the resistors have linear temperature coefficients only over limited temperature ranges. The same comment is often true for the zero coefficient, reference resistors. In addition to the possible nonlinearity of the resistor devices, many of the sensitive element assemblies have long thermal time-constants. Nevertheless, for special purposes and limited ranges, this device can be made to be quite accurate and satisfactory. Also, because both types of resistors are easily obtainable, such a circuit is simple and easily assembled.

Many modern systems use thermistors, which are essentially temperature-sensitive, semiconductor, resistance devices. These devices have better thermal time-constants than many of the more sensitive metallic resistor sensors, but the time-constant can be a problem. Also, many of these devices have limited upper temperature ranges.

Because the semiconductor materials, and especially their junctions, have high temperature coefficients, many of the downhole solid-state circuits monitor the temperatures by merely reading the voltage across one of the active circuit junctions, with a high impedance voltmeter circuit. This can give valuable clues to instrument operation and to borehole temperatures. This also means that temperature can be a problem with some types of semiconductor circuits.

Some of the circuits used to measure temperature are shown in Figure 6.3.

6.3.3 Temperature Measurement Considerations

Borehole temperature measurements have a number of limitations that should be considered. Thermal time-constant is a function of heat capacity and the thermal conductivity of the components. The thermal time-constant of the system is a function of the thermal time-constants of each of the components of the downhole instrument. These components include the sensor, the sensor housing, the stability of the rest of the circuitry, the surface circuitry, stability, linearity, and the mass of the instrument housing in the vicinity of the sensor.

A platinum resistor has a well-documented and linear temperature coefficient of resistance and can be used as a sensing resistor in a bridge circuit. They are bulky, however, and have a long time-constant.

In many cases, a platinum thermocouple couple can be placed directly in the fluid to be monitored. This is especially true in laboratory environments and in gases. Thus, thermocouples are extensively used in scientific laboratory measurements. Thermocouples, however, have low signal outputs and require a constant temperature reference.

The thermal time-constant is frequently a function of the protective housing used to insure reliable operation under pressure, corrosive environment, and in electrically conductive fluids. Thermistors, resistors and thermocouples, which themselves have short time-constants are frequently housed in

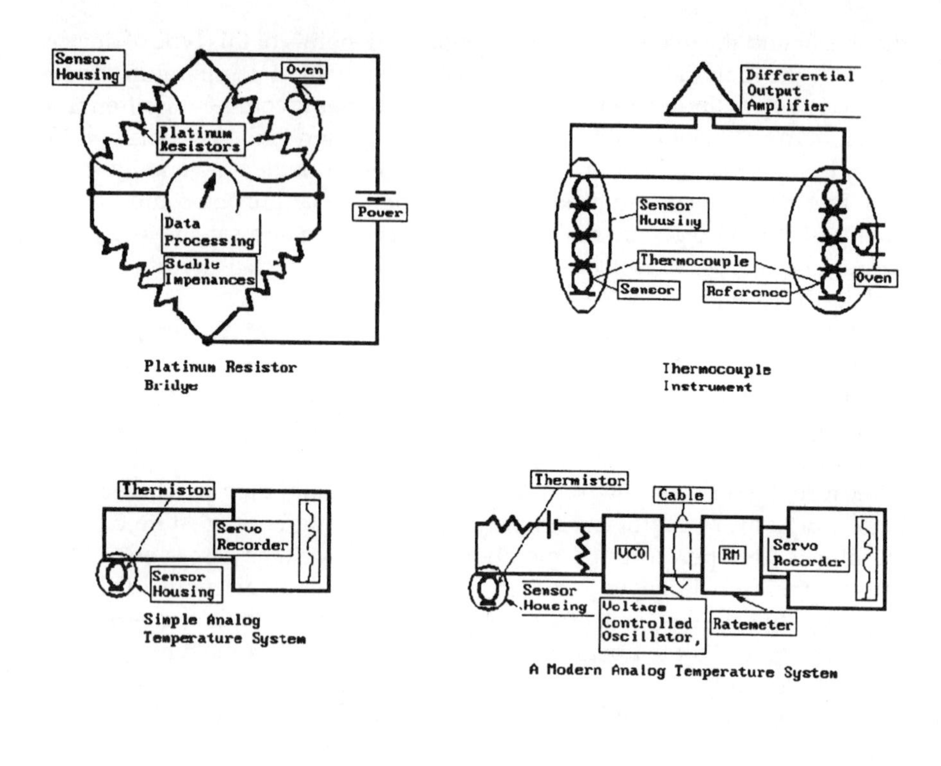

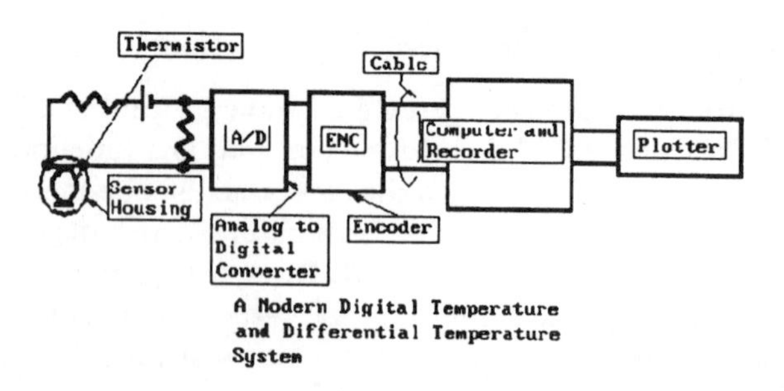

FIGURE 6.3
Various circuits for temperature systems.

stainless steel or brass pressure housings that extend the thermal time-constant by several magnitudes. This is because of the usual high heat capacity of housing metals. Also, for the high temperature measurements, magnesium oxide is a common insulating material. Unfortunately, it has a very low thermal conductivity, which may also lengthen the thermal time-constant.

The maximum logging speed depends upon the thermal time-constant. When the thermal time-constant is long, the logging speed is affected. In order to sense temperature variation details, it is necessary to log at low speeds. Five time-constants are required for the sensor to reach 98% of the temperature of the material outside. Thus, a 12-sec time-constant device will require an exposure of 1 min to reach 98% of the value being sought. A logging speed of 5 ft/min (1.5 m/min) will average the temperature reading over a hole length 5 feet (1.5 meters) for an accuracy of ±2% (refer to Figure 6.4). Thus, the temperature log is always a running average of the actual temperatures. The log's approach to reality will depend upon the net system thermal time-constant and the logging speed. The thermal time-constant must be as short as possible.

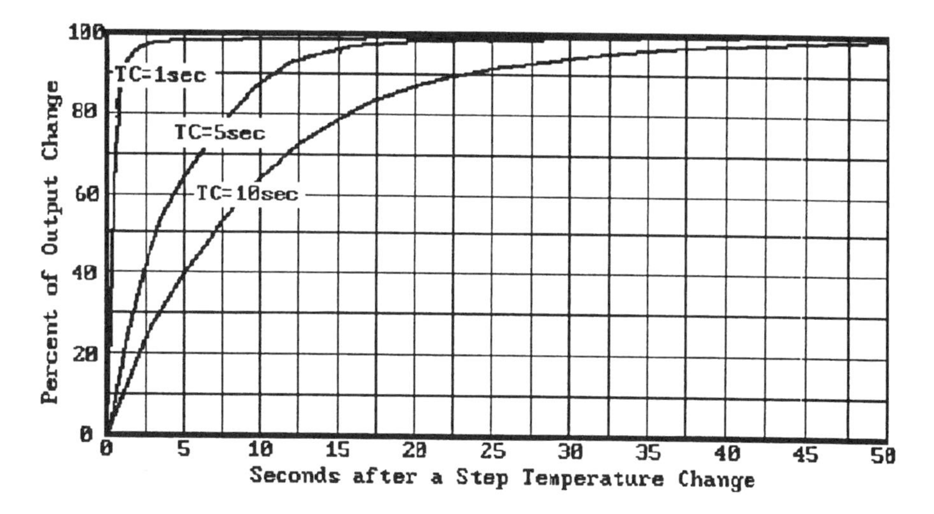

FIGURE 6.4
Effect of thermal time constants.

A metallic element resistor can be quite reliable and stable. Its resistivity is often low, however, so a long length of wire is needed to obtain a reasonable resistance for a bridge element. It must be protected from the borehole chemical and electrical environment by a housing. This is inconsequential in a stationery measurement. On the other hand, if a moving measurement is made the thermal time-constant of the total bulk of metal may severely limit the logging speed.

The design of the conduit for the flow of the fluid past and around the sensor is important. If the flow is restricted, this can also lengthen the time-constant. Large masses of metal, put in for mechanical protection, can also contribute substantially to the time-constant. Choose the device with care. Know what the thermal time-constant of that device is. Plan the measurements with the instrument characteristics in mind.

6.3.4 Differential Temperature Measurements

The measurement of the differential temperature (dT/dx) (the first derivative of the temperature gradient with respect to distance) is usually quite valuable, especially for detailed examination and for locating small changes.

Basically, dT/dx is determined by measuring the temperature at two places simultaneously and dividing the difference by the distance between the sensors. Mathematically, this is

$$\frac{\Delta T}{\Delta D} = \frac{T_2 - T_1}{h} \approx \frac{dT}{dx} \tag{6.5}$$

where

$\Delta T/\Delta D$ = the temperature gradient or the differential temperature.
D = the hole depth.
h = the distance between the two sensors.
T_1 and T_2 are the two temperature measurements.

Most modern equipment uses only one sensor. The temperature is memorized and compared with that measured a preset distance from the first measurement.

Some devices determine a time differential (dT/dt) rather than a distance differential. For locating small changes, this will allow a high degree of control over the sensitivity of differential temperature measurements. It should not, however, be confused with the temperature gradient measurement, dT/dx. It is necessary to know the relationship between the indicated depth differences on the log and the logging speed over that interval to convert this measurement to the temperature gradient. This conversion is

$$\frac{\Delta T}{\Delta x} = \frac{\Delta T}{\Delta t} \frac{\Delta t}{\Delta x} \tag{6.6}$$

where

$\Delta T/\Delta x$ = the temperature gradient with respect to distance.
$\Delta T/\Delta t$ = the temperature gradient with respect to time.
$\Delta t/\Delta x$ = the reciprocal of the logging speed in the same units as the gradients.

Exxon experimented (1979) with a rotating differential measurement to detect temperature anomalies on a plane normal to the axis of the borehole. The purpose of the system was to detect temperature anomalies that might be due to cavities in the cement annulus between the casing and the formation. Location of the cavities allowed oriented perforating and cement squeezing to eliminate the voids. Exxon reported as much as 3°F, in some cases, from one side of the hole to the other. As far as this author is aware, the method was not used beyond these experiments.

6.4 Uses of Temperature Measurements — Correction of Solution Resistivities

The most common use of temperature measurements in the borehole is to allow correction of the solution resistivities. The electrical conductivity of an electrolyte is a function of its temperature. As temperature increases, the viscosity of water decreases, thus the mobility of the ions increases. The relationships are

$$
\begin{aligned}
C_W &= 1+\alpha_{^\circ C}\left(^\circ C - 18^\circ\right)C_{18} \\
C_W &= 1+\alpha_{^\circ F}\left(^\circ F - 64.4^\circ\right)C_{64.4} \\
R_W &= \frac{R_{18}}{1+\alpha_{^\circ C}\left(^\circ C - 18^\circ\right)} \\
R_W &= \frac{R_{64.4}}{1+\alpha_{^\circ F}\left(^\circ F - 64.4\right)}
\end{aligned}
\tag{6.7}
$$

where

C_w = the determined solution (water) electrical conductivity.
R_w = the determined solution electrical resistivity.
C_{18} = the electrical conductivity of the solution at 18°C (64.4°F).
R_{18} = the electrical resistivity of the solution at 18°C (64.4°F).
α_T = 0.025 Ωm/°C = 0.014 Ωm/°F, the temperature coefficient of resistivity of the solution.
A_T = $R_w/R_{18} = 1/[1 + \alpha_{^\circ C}(^\circ C - 18^\circ)] = 1/[1 + \alpha_{^\circ F}(^\circ F - 64.4^\circ)]$.

The U.S. oil field operations have, for many years, used the relationship

$$
R_{W2} = R_{W1}\frac{^\circ C_1 + 21.539^\circ}{^\circ C_2 + 21.539^\circ} = R_{W1}\frac{^\circ F_1 + 6.77^\circ}{^\circ F_2 + 6.77^\circ} \tag{6.8}
$$

Because the calculation is an approximation and not highly accurate, for many years, this author has used the value of 7°F instead of 6.77°C (and 22°C instead of 21.539°C). This is discussed in Chapters 3 and 5 in *The Introduction to Geophysical Formation Evaluation*. Another method used in the oilfield for finding the probable average temperature gradient of the borehole fluid is shown in Figure 6.5.

The value of R_{W2}, found with Equation 6.8, is the approximate value of the resistivity of a dilute solution of NaCl *in the borehole at the time of logging*. This procedure is necessary, because many of the solutions used have their resistivity measured at some temperature other than that at the formation level.

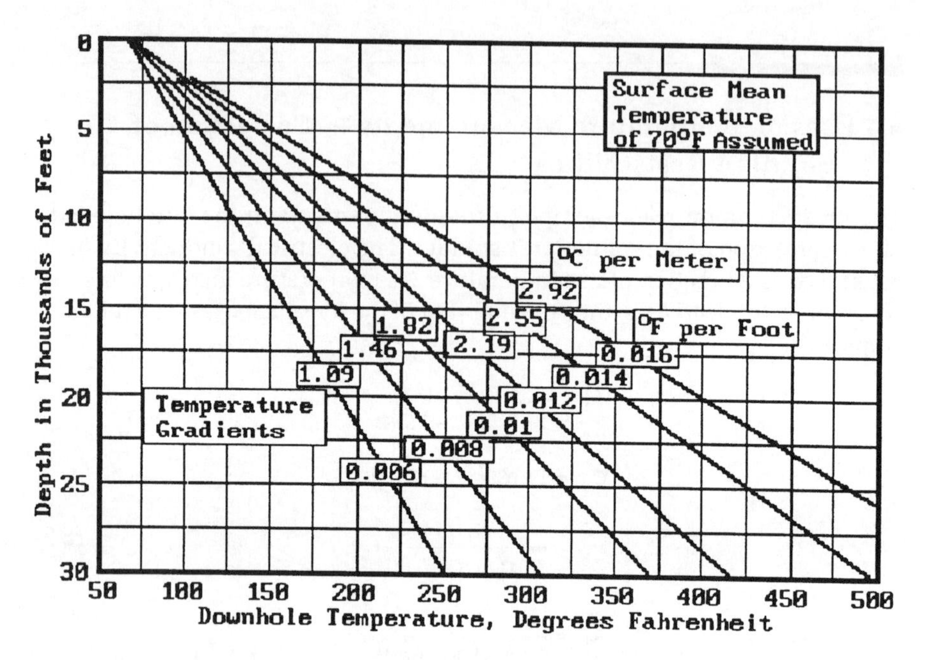

FIGURE 6.5
The standard method of estimating probable borehole temperatures.

Note that this temperature *IS NOT THE FORMATION TEMPERATURE.* It is the temperature of the borehole fluid column at the formation level at the time of measurement. It is found by determining the *average* temperature gradient of the borehole fluid, estimating the temperature of the borehole fluid at the formation level, estimating the composition of the mud or water, and assuming that the mud or water has the same temperature characteristics as a dilute solution of NaCl. This is an approximation at best.

A better method, of course, is to measure the temperature and solution resistivity at the formation level at the time the other logs are run. This is a service offered by most logging contractors. An alternative is to measure the mud resistivity at the formation level and compare it with that measured at surface temperature. Then, Equation 6.8 may be used to determine the *probable* temperature at the formation level.

6.5 Thermal Conductivity

The *in situ* measurement of the thermal conductivity of the earth formation zones has been (mostly experimentally) mainly done for scientific studies. The service is apparently not offered commercially at the time of this writing.

This can be easily done, however, in the laboratory. Several methods have been tried. The usual method is to apply a known quantity of heat energy and to measure the rate of temperature change at a distance from the source. This has been done regularly in the laboratory and in the borehole. It is also feasible to estimate the relative thermal conductivity by injecting water at a different temperature than ambient temperature, and measuring the time it takes to return to ambient temperature.

Thermal conductivity differences of the various zones of a formation will distort the shape of the isothermal distribution in the formation. This is because a thermal field exists from the center of the earth to outer space. The rate of flow of heat outward in each component from the earth's center will depend upon the thermal conductivity of each zonal component. This is illustrated in the simple model of Figure 6.6. Table 6.2 is a chart of thermal conductivities.

6.6 Determination of the Formation Temperature

It is important to be aware of the difference between the temperature that is routinely measured in the borehole and the true formation temperature. Actual formation temperature is important because the undisturbed formation is where the ore is. Knowing the true formation temperature will allow a better assessment of the other formation parameters.

Temperature measurements made in the borehole during logging can be used to estimate the probable formation temperature. As soon as drilling and circulation stops, the heat imbalance of the borehole and adjoining formation will begin to move to rectify the imbalance. Temperature measurements made over frequent intervals will show a change in value toward the formation temperature. Because the rate of flow of the heat will depend upon the difference of temperature, the temperature change will be logarithmic, not linear. The curve of the temperatures vs. time of the borehole fluid will, however, be asymptotic to the formation temperature.

One method that is frequently used to determine the formation temperature is to place disposable temperature sensors at known depths in the borehole and allow the well to be cased, filled, and/or abandoned. Temperature measurements are made at intervals after work has stopped. When the temperatures cease to change, the thermal environment can be said to have become fully stabilized.

Another method that is used is to make several temperature measurements over as long a time interval as possible during a single drilling break. The temperatures are plotted against elapsed time. Frequently, the curve can be found to be asymptotic to a temperature value. This is presumed to be the formation temperature.

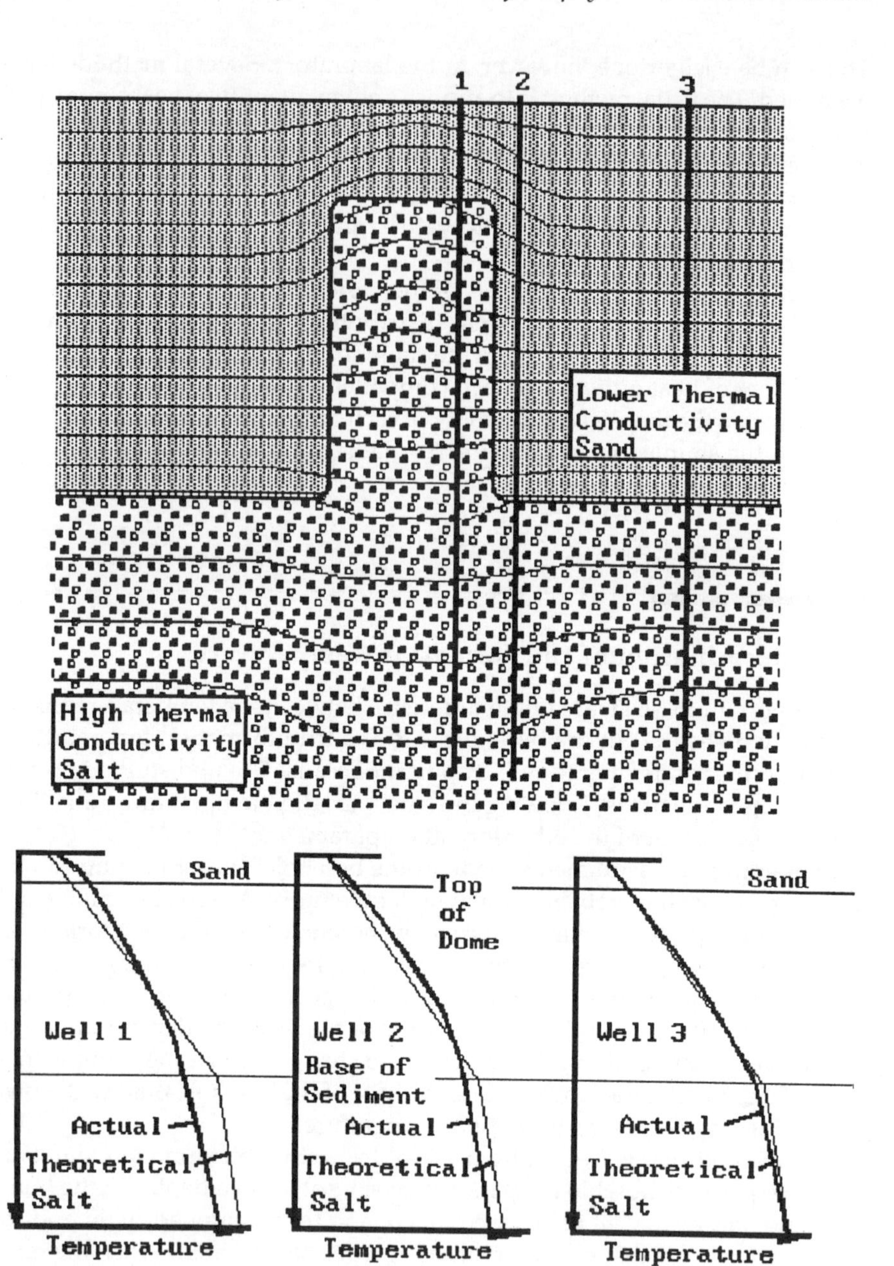

FIGURE 6.6
Thermal distortions caused by thermal conductivity anomalies. (After Guyod.)

Fertl and Wichmann (1977) suggest plotting the time ratio on a semilogarithmic scale, where t = the time circulation was stopped and Δt is the elapsed time after circulation was stopped. Then,

TABLE 6.2

A Table of Thermal Conductivities of a Few Minerals

Thermal Material	Conductivity 10^{-3}(cal/cm² sec)	Temp. (°F)	Thermal Material	Conductivity 10^{-3}(cal/cm² sec)	Temp. (°F)
Carbon dioxide	0.04000	80	Aluminum	530	
Carbon dioxide	0.044	120	Copper	940	
Ethane	0.043	32	Shale	2-4	
Ethane	0.051	80	Sandstone	3-12	
Ethane	0.074	200	Fused silica	3.2	
Air (dry)	0.057	32	Limestone	2.4-8	
Air (dry)	0.061	68	Polyhalite	3.7	
Air (dry)	0.074	212	Serpentine	4.3-5.9	
Air (dry)	0.088	392	Basalt	4-7	
Nitrogen	0.062	80	Granite	5-8.4	
Nitrogen	0.065	120	Calcite	5.13	
Methane	0.073	32	Feldspar	5.8	
Methane	0.081	80	Slate	6	
Methane	0.106	200	Norite	6.42	
Crude oil	0.3	68	Granopyrite	6.64	
Helium	0.332	68	Quartz	6-30	
Kerosene	0.357	86	Quartz C-axis	26	100
Sulfur, monocl.	0.38	212	Quartz C-axis	22	200
Sulfur, rhombic	0.56	176	Quartz A-axis	14	100
Sulfur, rhombic	0.65	68	Quartz A-axis	12	200
Coal, lignite	0.33	68	Syenite	7.66	
Portland cement	0.71	68	Halite	14.3 (8-15)	
Water	1.39	32	Garnite	8.5	
Water	1.43	68	Dolomite	9.3-11.9	
Water	1.60	167	Dunite	10	
Clay	2-3		Chlorite	12.5	
Chalk	2-3		Anhydrite	7-13.4	
Gypsum	2-4		Quartzite	16.05	
Lead	83		Pyrite	25-40	
Iron	180	68	Hematite	25	
Magnesium	380	68	Magnetite	30	
			Sphalerite	63.6	

$$\frac{\Delta t}{t + \Delta t} \tag{6.9a}$$

$$\frac{\Delta t}{t + \Delta t} \rightarrow 1 \quad \Delta t \rightarrow \infty \tag{6.9b}$$

When $\Delta t = \infty$, the time-ratio becomes equal to 1 and its logarithm will be equal to 0. The indicated temperature will then be equal to the static or formation temperature. This is shown in Figure 6.7. Plot the temperature on the

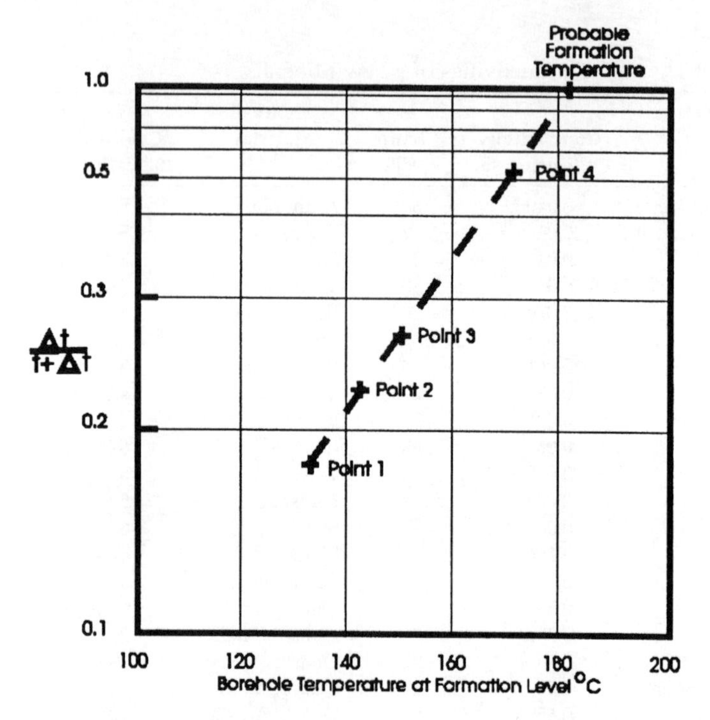

FIGURE 6.7
The Fertl-Wichmann method of estimating the probable formation temperature. (Courtesy of Atlas Wireline, Inc.)

linear scale and the time function on the logarithmic scale. The equation for this is

$$Slope = \frac{T_2 - T_1}{\log t_2 - t_1} \tag{6.10a}$$

where

T_n = the measured temperature at any time.
t_0 = the dimensionless time, y-intercept, when the logarithm of the dimensionless time is equal to 0.
t_n = the dimensionless time, $\Delta t/(t+\Delta t)$, corresponding to T_n.
T_1, T_2 = borehole temperature measurements.
t_1, t_2 = are the dimensionless time parameters corresponding to T_1 and T_2, respectively.

This expression is easily illustrated by plotting, as shown in Figure 6.7. Extrapolate the trend of the temperatures to a value of the ratio equal to 1.0. The logarithm of the ratio, at that time, will be zero. Thus, T_n will indicate the probable borehole temperature at an infinite time after circulation has stopped.

Several years ago (c1976), the IRT Corporation field tested a logging system to measure the "formation temperature". This device determined the average energy content of the thermal neutrons within the formation. This was equated to the "formation temperature". Because the peak of the population of thermal neutrons is rather shallow (6 to 18 in.; 15 to 45 cm radius from the source, depending upon the formation porosity), the measurement is not always a correct value of the *undisturbed* formation temperature; but it is more likely to be that of the invaded zone. This is, however, a step nearer to the true formation temperature.

6.7 Injection of Fluids Into the Formation

Petroleum wells, injection wells, and *in situ* mining (solution mining) holes can make use of temperature gradient and differential temperature logs to determine the location of solution entry into the formation and its distribution. The injected fluid should be at some other temperature than the formation temperature. This is usually no problem because the surface water is usually cooler than the formation water. This injected water will be cooled (or warmed) by the permeable formation and will change the temperature of the borehole fluid.

Figure 6.8 shows a log which was run in a cased well at 3100 to 3500 ft (945 to 1067 m). This situation, or a similar one, could occur equally well in a water well or a disposal well. The well had been shut in for 63 hr after the fluids had been injected into the formation through perforations in the casing. The logs are temperature gradient, differential temperature, gamma ray, injection temperature gradient, and a time travel tracer log. Note that the temperature logs show that almost all of the injected fluid was taken by the perforations between 3228 and 3560 ft. No fluid appears to have entered the formation below 3390 ft.

Figure 6.9 shows a log run in an injection well that had slotted perforation liners. This also could be a commercial disposal well. The injection rate was 2600 bbl/d at 800 psi for 3 years. The well was shut in for 30 hr before logging. A temperature gradient, differential temperature log, and a radioactive tracer profile log were run. There was good agreement between the temperature logs and the tracer profile. Note that the normal earth temperature gradient has been completely altered by the long term injection. The top two perforations took the bulk of the injected water (65%) and a substantial portion went into the lowest one third of the bottom slot.

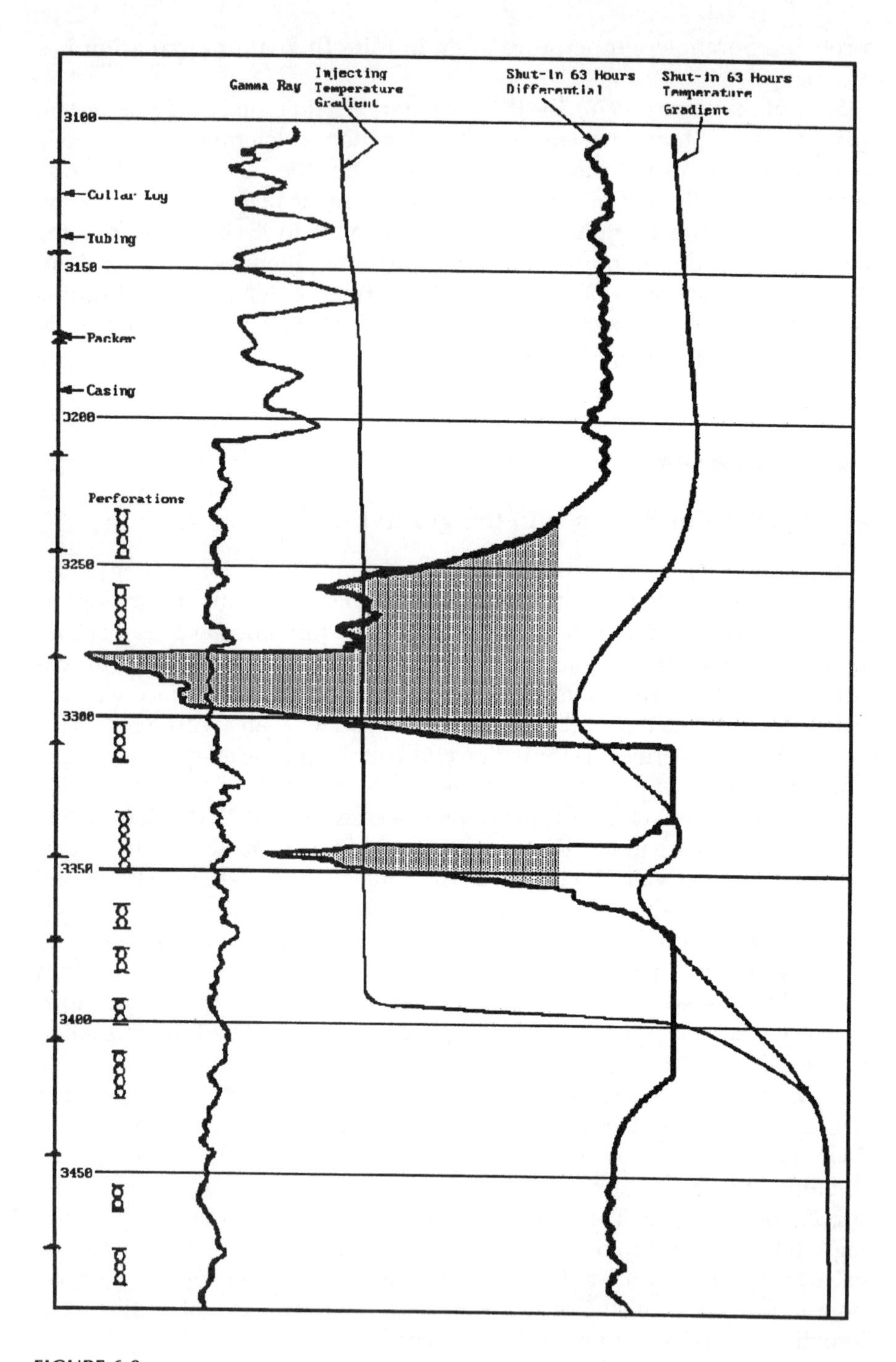

FIGURE 6.8
A cased hole temperature log showing the effect of fluid injection.

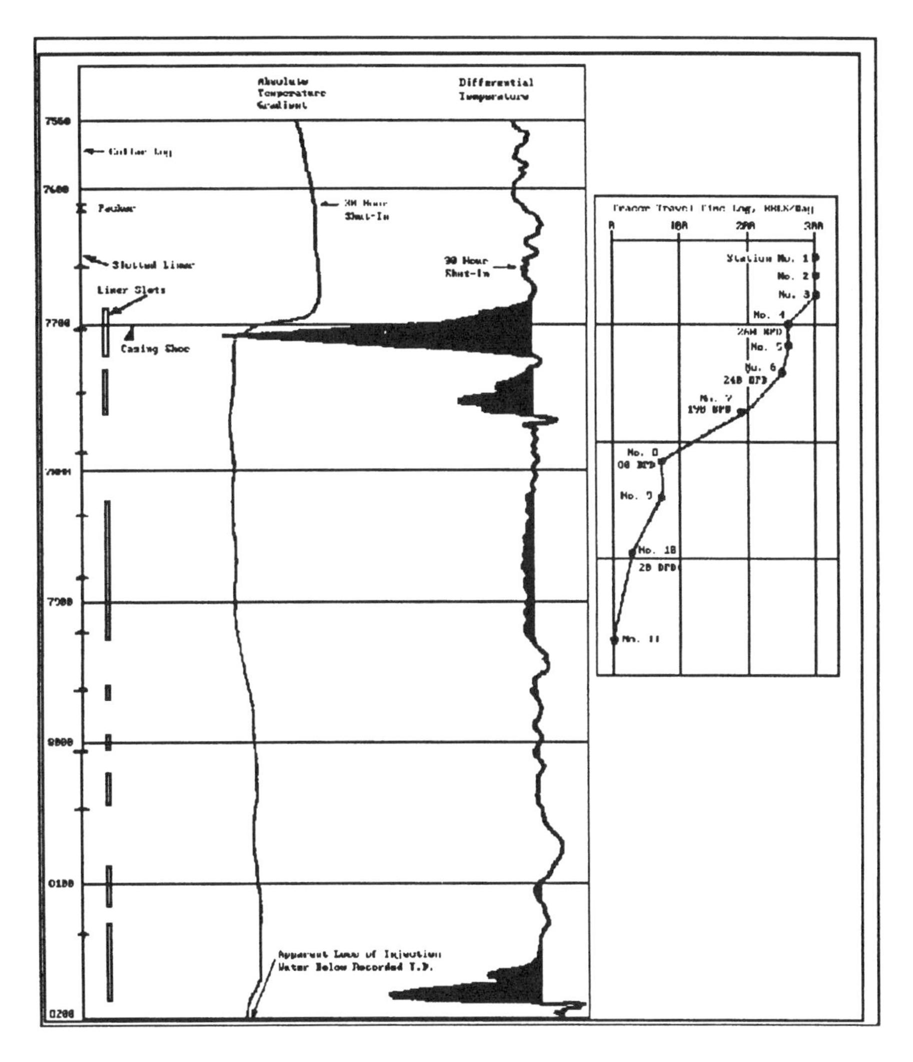

FIGURE 6.9
A temperature log in an injection well having a slotted liner.

6.8 Fluid Moving Into and Out of the Borehole

If the formation fluid enters the borehole, the temperature of that fluid is seldom at the temperature of the borehole fluid, unless the hole has been undisturbed for many days. Thus, if formation fluid entry is suspected, temperature anomalies should be investigated. This is especially true if the anomaly is in the same direction as the probable formation temperature (i.e., warmer or cooler than the borehole temperature).

Figure 6.10 shows a normal water entry log. The well had been shut in for 24 hr when the first run was made. The log shows a nearly normal temperature gradient. The second temperature log was then run with the well flowing. The temperature gradient log clearly shows warm formation water entering at the perforations at 6266 to 6271 ft (1910 to 1912 m). The different gradient above the perforations verifies the production of formation fluids. The coincidence of the two gradient measurements below the perforations indicates that no fluids are entering from below.

In the development of water wells, it is common practice to run a temperature log with the drilling mud at normal pressure. Then, the pressure is deliberately lowered and another temperature log is made. New temperature anomalies indicate possible water production.

After casing has been set in a gas well, and perforations have been made, the production of each level may be determined by allowing the well to flow during the logging of the well temperature and differential temperature. The adiabatic cooling of the entering gas will produce lower temperature anomalies which will be located in the reservoir at the gas entry point. This can sometimes be a problem in sedimentary uranium exploration boreholes. A log of a well in such a situation is shown in Figure 6.11. The log shows substantial cooling in the sand above the perforations. The rate at which the slope of the temperature curve returns to a normal gradient is proportional to the amount of gas entering. This is shown in Figure 6.12. The gas production rate, GPR, in millions of cubic feet per day, is

$$\Delta t = T_0 - T_g \frac{dD}{dT_g} \tag{6.10b}$$

where

T_0 = the undisturbed borehole temperature at the zone of interest.
T_g = the actual temperature of the gas at that zone.
dD/dT_g = the reciprocal change of temperature of the gas with depth (minus the slope of the temperature curve).

T_0 is determined by drawing the normal gradient through the anomaly. It is the temperature that would be read if no gas were entering.

A tubing or casing leak will often put borehole fluid into the annulus or formation. This can be a serious problem in disposal, water, and/or monitoring wells. If the borehole fluid is at a different temperature than the formation, this problem can be detected with the temperature/differential temperature log. The extra fluid, at a different temperature, will cause a temperature anomaly on the gradient curve. This will be outstanding on the differential temperature curve. The log of Figure 6.13 was run because a problem was suspected. This was a casing leak that put borehole fluid into the annulus.

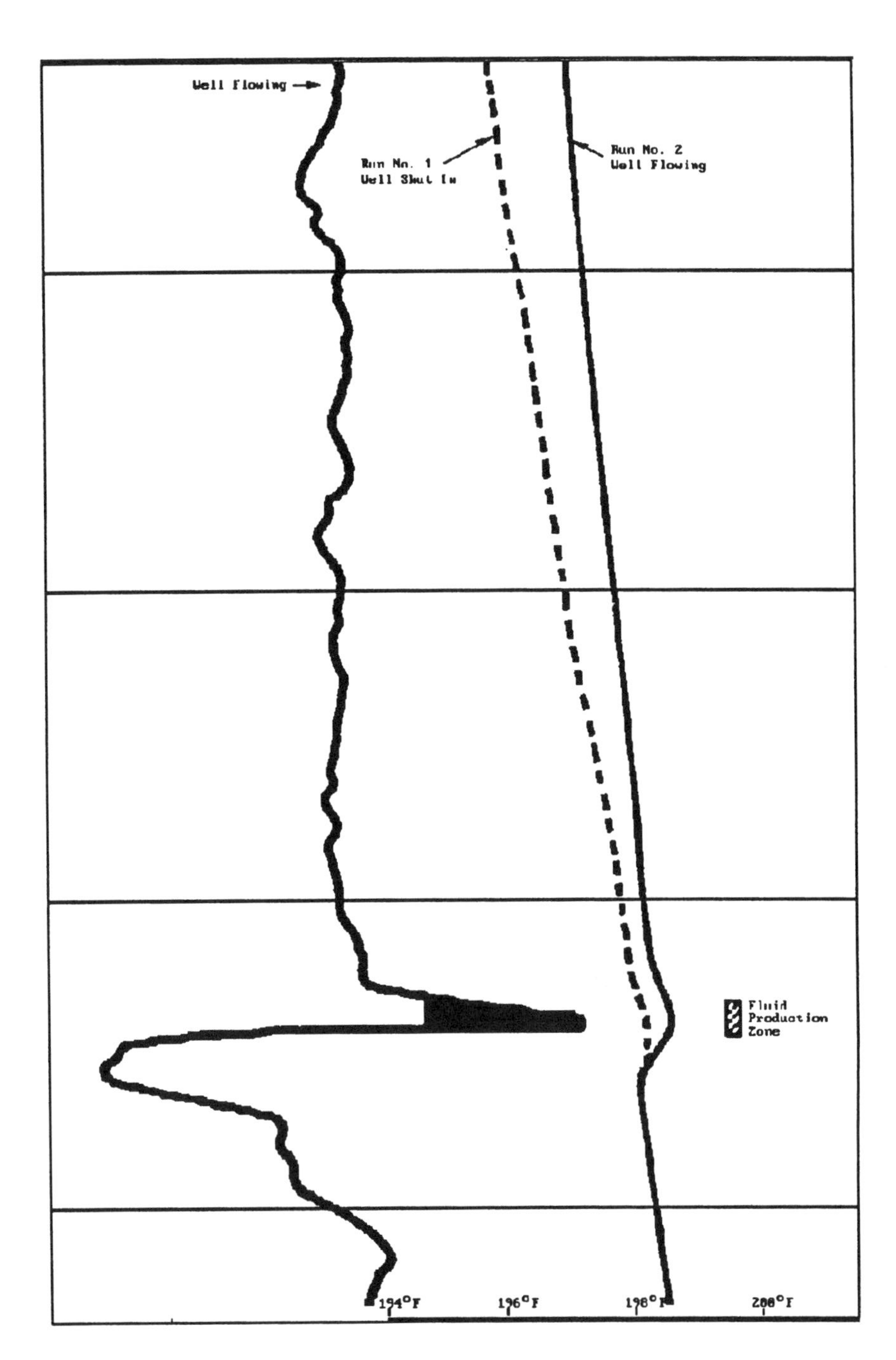

FIGURE 6.10
Temperature log showing water entry. Initial shut-in and flowing.

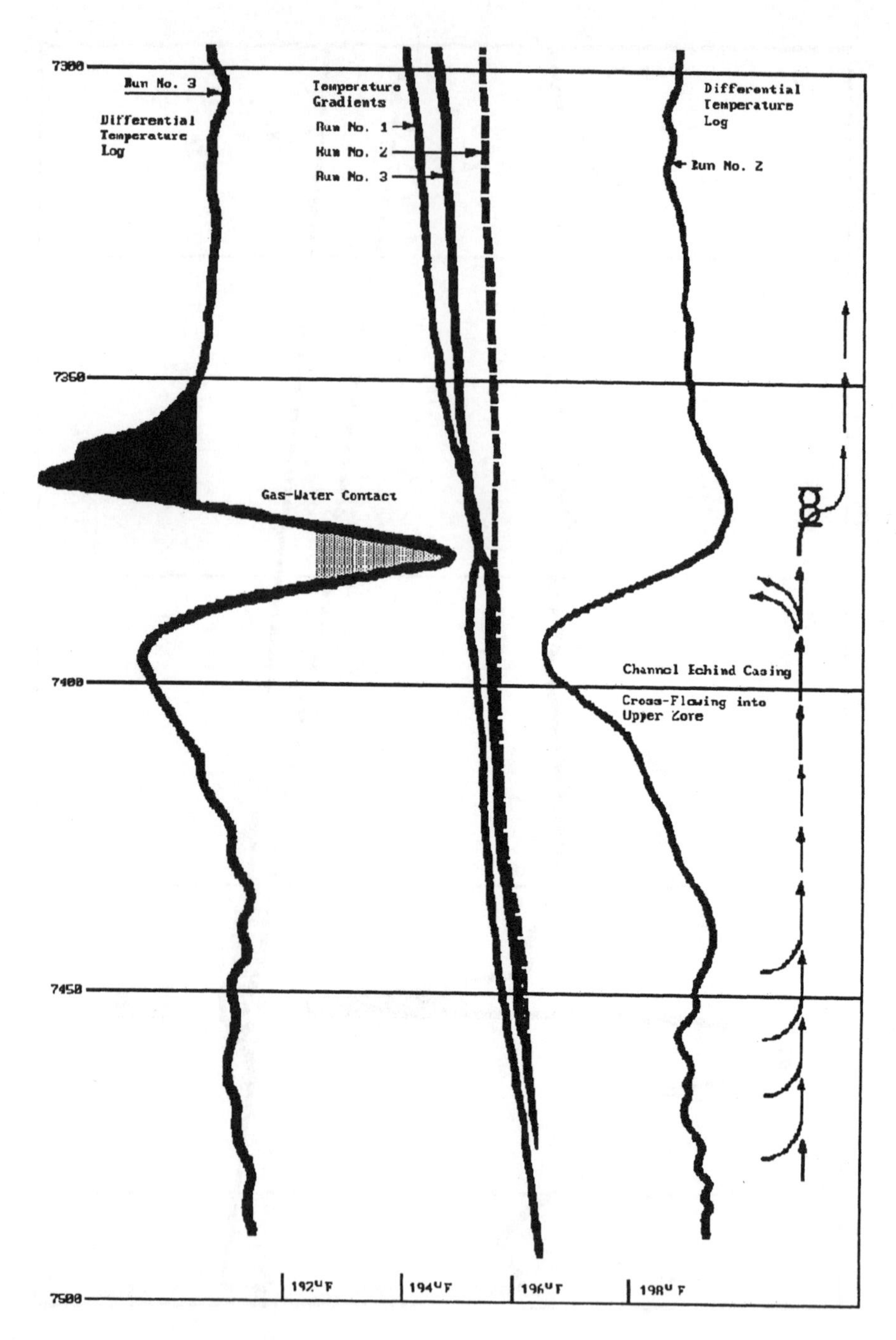

FIGURE 6.11
A log showing gas/water contact.

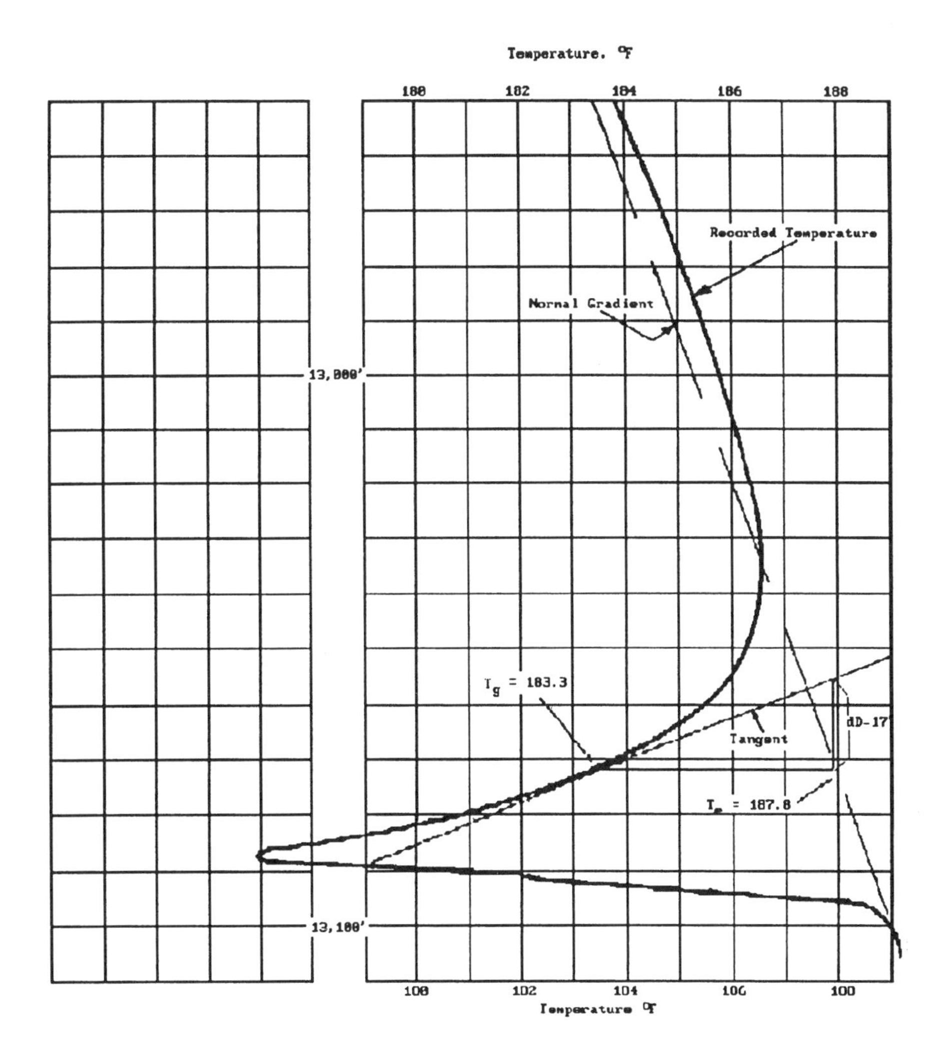

FIGURE 6.12
A diagram depicting the method of estimating the amount of gas entering.

6.9 Fluid Loss and Communication

Temperature profiling can be used to find fluid loss in cased holes and to locate communication between a perforated zone and the adjoining zones, a potential problem in water and solution mining wells.

If fluid loss or a leaking casing is suspected, water, at a temperature different from the borehole or formation temperature, can be pumped into the casing immediately after a normal temperature and differential temperature log has been made. Often it is feasible to use the drill pipe to spot the hot or cold

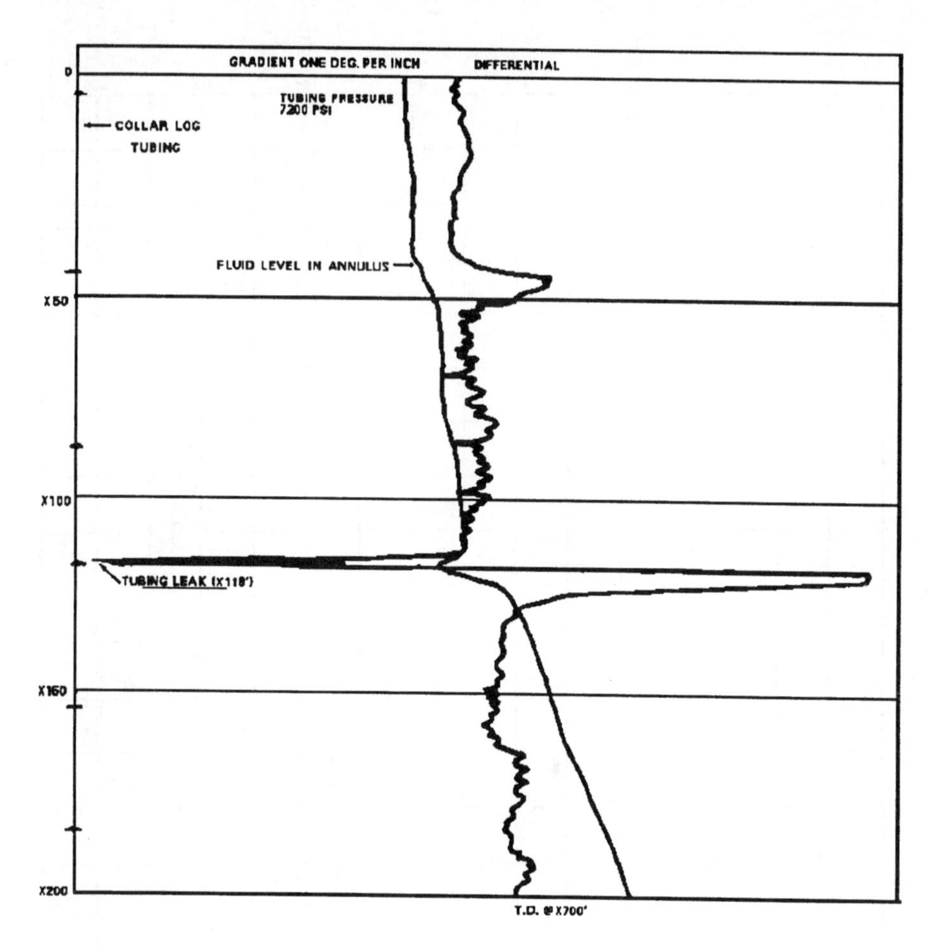

FIGURE 6.13
Log shows a casing leak.

water at the suspected fault zone. Then, temperature logs at regular intervals will record the thermal change with time. If the casing has leaked, the thermal change at and near the leak will be slower than in the intact zone of the casing.

If the same operation is performed on a perforated zone, the permeable formation zone behind the perforations will take large amounts of the water. The maximum temperature anomaly will appear at the perforations and may extend to the limits of the permeable zone, if the perforations are extensive. The rate of temperature change, with time, will indicate the amount of water taken, and thus, the relative permeability of the zone. Any communication through the annular cement will appear as a lower amplitude anomaly, which will die away faster than the permeable zone anomaly. Communication with another permeable zone will appear as another anomaly, similar to the main zone anomaly, but probably of lower amplitude. It will die away

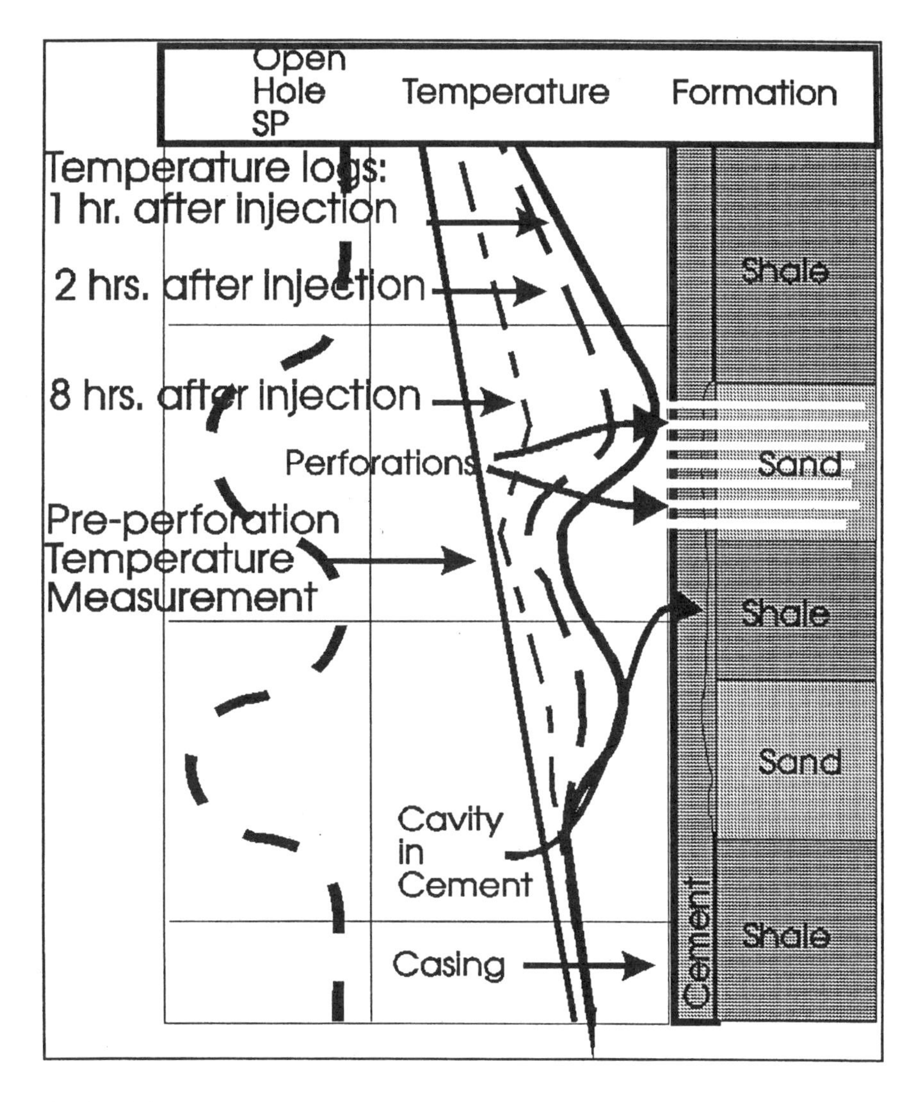

FIGURE 6.14
A hypothetical log showing annular cement leakage.

more slowly than the cement channel anomaly. Figure 6.14 shows a hypothetical temperature profile log made over 24 hr.

If channeling occurs behind the casing, the formation fluids from one zone can flow to another, lower pressure zone. This can happen during production and, also, if there are holes in the casing. The result will often be the distortion of the normal temperature curve within the well. Figure 6.15 shows a temperature log where gas has infiltrated the lower fluids. This is a possible problem in deep commercial water wells, as well as oil/gas wells.

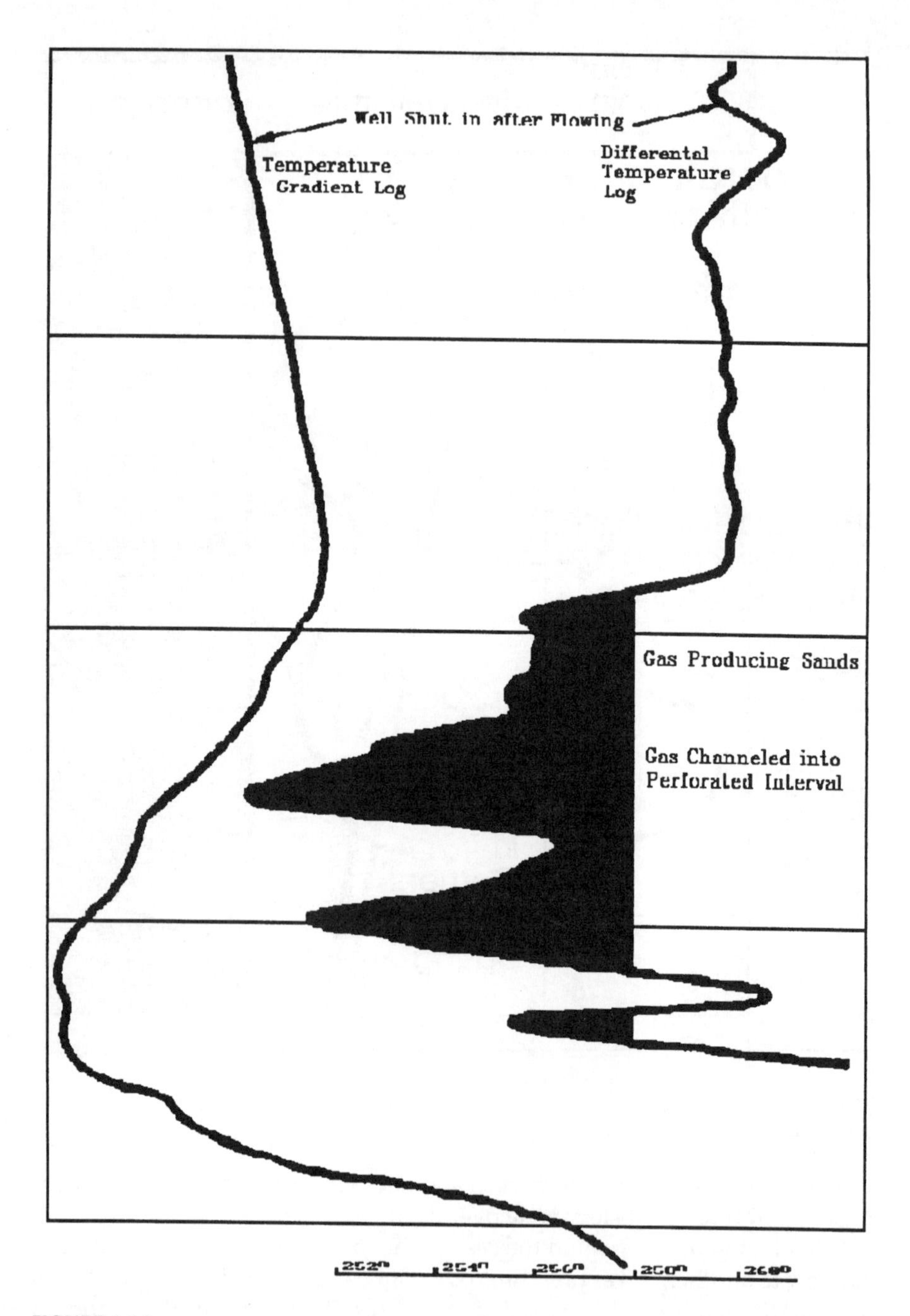

FIGURE 6.15
Gas infiltrating lower sands.

These are but a few of the possible applications of these systems. Many more are cited in the literature and possibilities are limitless.

7

Borehole Calipers

The borehole caliper is one of the early diagnostic systems. Early calipers generally consisted of "go-no go" devices. The electrical cable allowed real diameter measurements to be read out on the surface in real time. Caliper devices are used in both cased and open holes. There have been at least three different types of devices used.

The first commercial contractor's wireline, surface-readout caliper was Schlumberger's "Section Gauge". It had three steel bowsprings anchored to a brass mandrel at the upper end and moving a plunger on the lower. The plunger consisted of ferrite beads and brass washers whose sizes and spacings compensated for the nonlinear motion (with respect to hole diameter) imparted to it by the motion of the bowsprings. The plunger moved within a differential transformer coil. The output of the transformer depended upon the position of the plunger within the coil and was a linear function of borehole diameter.

Later calipers used three hinged arms which operated a single linear potentiometer. Calipers were also added to sidewall devices, such as the MicroLog, the micro-resistivity tools, the sidewall neutron, the sidewall acoustic, and the density systems. Since then, calipers have been incorporated into many of the downhole systems. Geophysical logging calipers have been offered that have from one to six arms. Mechanical diagnostic calipers may have as many as 64 arms. Acoustic calipers are available. Downhole digitizing has replaced analog transmission on many systems. Display systems enhance the ease with which the downhole environment can be visualized. Many systems display the caliper signals in real time and use them to make necessary corrections to other signals.

7.1 Single-Arm Devices

The single-arm devices are to be found incorporated in the sidewall devices: the microresistivity tools, density, sidewall acoustic, and sidewall neutron. They are designed primarily to determine the presence of mudcake and its

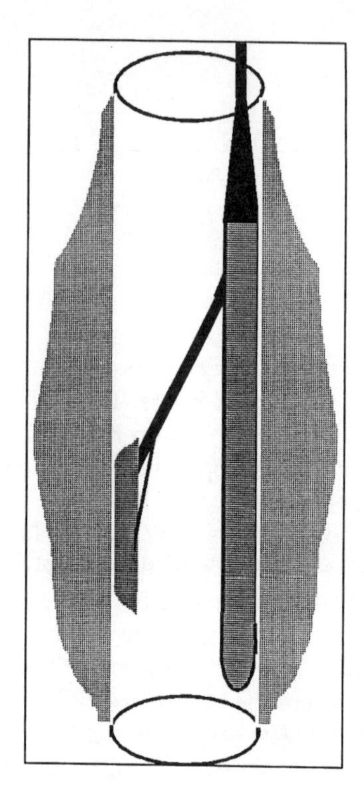

FIGURE 7.1
A single-arm caliper on a density tool.

thickness on the wall of open holes (see Figure 7.1). *Single-arm calipers are not intended for borehole diameter use and are inaccurate for that purpose.* These devices are designed to ride on the surface of the mudcake and do not read the true diameter of even a circular cross-section borehole. Also, few borehole intervals are truly circular. Single-armed devices will tend to ride the major diameter of an out-of-round hole.

Mudcake character (thickness, resistivity, density, transmission time, water content, trace element content) information is needed to make corrections for the presence of a distorting mudcake signal component. Mudcake thickness is also a good indicator of the quality of the drilling mud and invasion. This is especially important with sidewall logging devices.

Mudcake thickness, T_{mc}, is usually assumed to be

$$T_{mc} = \frac{d_c - d_b}{2} \tag{7.1}$$

where

d_c = caliper reading.
d_b = the drill bit diameter.

The value of T_{mc} becomes important with the microresistivity methods because the volume of investigation of any of the microresistivity devices is small. A mudcake between the electrode array or transmitter and the formation can form a substantial part of the measured volume. The mudcake occurs in what is usually the most sensitive region of the measurement volume. Thus, to get an accurate value for the near-borehole resistivities, R_i, and R_{xo}, the contributions of the mudcake resistivity must be determined and eliminated from the total reading. Logging tool manufacturers and logging contractors publish empirical charts for manual analog correction for mudcake effects. Many modern digital systems make the mudcake correction automatically, often in real time.

The resistivity of the mudcake can be determined on the surface. Because the measurement geometry and the sensitivity distribution are known, only the mudcake thickness must be determined. Then the contribution to the total signal of the mudcake can be determined and eliminated from the total signal.

The same reasoning is used with regard to the mudcake corrections to the total signals of the other sidewall devices. Usually, the affecting parameters for these other devices have a much smaller range of values than does resistivity. Mudcake resistivity can vary from about 0.05 to 5.0 Ωm, a range of 10,000%. Mudcake densities vary only from about 1.0 to 1.5 g/cm^3, a range of 50%. Water content and acoustic travel times have ranges of variation similar to that of mudcake densities. Thus, the corrections for these nonmicroresistivity devices are generally less critical and exacting than those for the mudcake resistivity. Many of the "borehole-compensated" tools will make corrections automatically.

7.2 Two-Arm Calipers

Two-arm calipers are in reality, variations of the single-arm devices. Many of the sidewall tools, such as the MicroLog, are designed with two arms and sensors on both. The comments of the single-arm devices apply equally to the two-arm calipers (see Figure 7.2).

7.3 Three-Arm Calipers

Three-arm calipers have arms at 120 degrees around the downhole tool, and the arms are linked. That is, they move together. This type of device gives a better picture of the hole diameter because it is much less affected by the

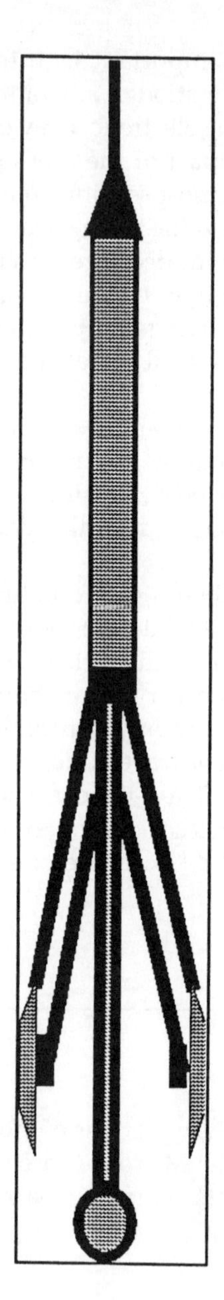

FIGURE 7.2
A two-arm caliper on a micro-resistivity sonde.

cross-section shape of the hole than are the single- and two-arm tools. It assumes, mechanically, that the cross section is circular (see Figure 7.3).

The original Schlumberger Section Gauge, all of the other early, single purpose calipers, calipers used for cementing, and early cased hole calipers were of the three-arm type. For purposes of determining average hole diameter and volume, they were far superior to the single- and two-arm types. They

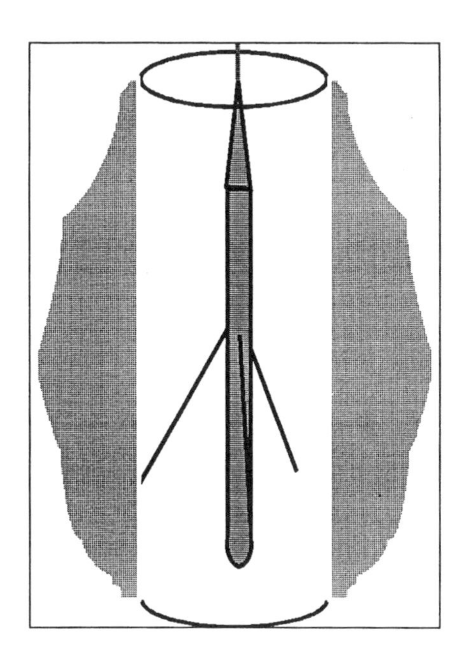

FIGURE 7.3
A three-arm dedicated caliper.

suffer, however, from the fact that they do not take the shape of the cross section into account.

7.4 Four- and Six-Arm Calipers

The original four-arm, open hole calipers were found on the first high-resolution dipmeter sondes. Because the four arms were linked, they were of no greater improvement as far as hole condition determinations were concerned than the three-arm systems. Their purpose was to supply redundancy to the dipmeter readings, in case one pad left the wall of the hole and faulty readings resulted. Later versions linked only opposite arms and were actual dual caliper systems. Thus, two caliper signals were generated: one, presumably, of the major diameter and the other of the minor diameter. The dual caliper methods were a major improvement over the earlier systems and raised the caliper to the full diagnostic tool category.

Six-arm calipers are being used to determine, more accurately, the actual cross-section profile of the borehole. The four- and six-arm calipers have opposite arms linked and each pair is independent. Thus, the four-arm caliper gives two curves which show diameters at 90 degrees to each other. The six-arm caliper makes three diameter readings at 60 degrees to each other.

These can be independent tools, but are most commonly seen on the dipmeter systems. Most dipmeter sondes now incorporate multicurve calipers.

7.5 Multi-Arm Calipers

Various dedicated caliper tools are available that have more than two independent pairs of arms. Gearhart Owens, Inc. built a six-arm caliper that had three independent pairs. This, of course, gave a somewhat better picture of the hole cross section.

Many smaller, multi-arm caliper have been used for cased hole work. These have many independent arms (as many as 64). They are used to locate and analyze casing corrosion, collapse, splits, and breaks. These have been partially replaced by the television and televiewer services. Figure 7.4 shows a log made with a 16-arm caliper in a 3.5 in. (8.9 cm) inside diameter tubing. Figure 7.5 is a copy of a multi-arm caliper log showing a hole in the casing. This particular incident was verified, later, by examining the hole in the recovered casing. This type of casing fault can be devastating in a water well or a disposal well.

7.6 Nonmechanical Methods

Recently, nonmechanical methods have been adapted to caliper use. Most of these have been acoustical methods. See Acoustic Methods (Chapter 25) and Radiation Methods (Chapter 22).

The propagation of a mechanical (acoustical) pulse through the borehole liquid can vary due to different densities and viscosities of the fluid. Modern acoustical tools, such as the borehole acoustic televiewer (BATV), often incorporate a separate system that sends a pulse only through the borehole liquid. The response time of this pulse is then used to correct the times of the measured signal, such as from the caliper device. The transit time from the transducer to the wall of the hole and back is easily measured and interpreted as the diameter of the borehole, casing, or tubing. Figure 7.6 shows the basic principle of the acoustic caliper.

The usual form of the acoustic caliper uses a rotor containing one or more transmitting and receiving transducers. The rotor spins in a plane normal to the axis of the hole. It is usually enclosed in an oil-filled housing at borehole pressure. As it rotates, it radiates a series of short, high frequency pulses. The pulses are reflected from the wall of the hole and sensed by the receiving transducer (which may or may not be the transmitting transducer). The

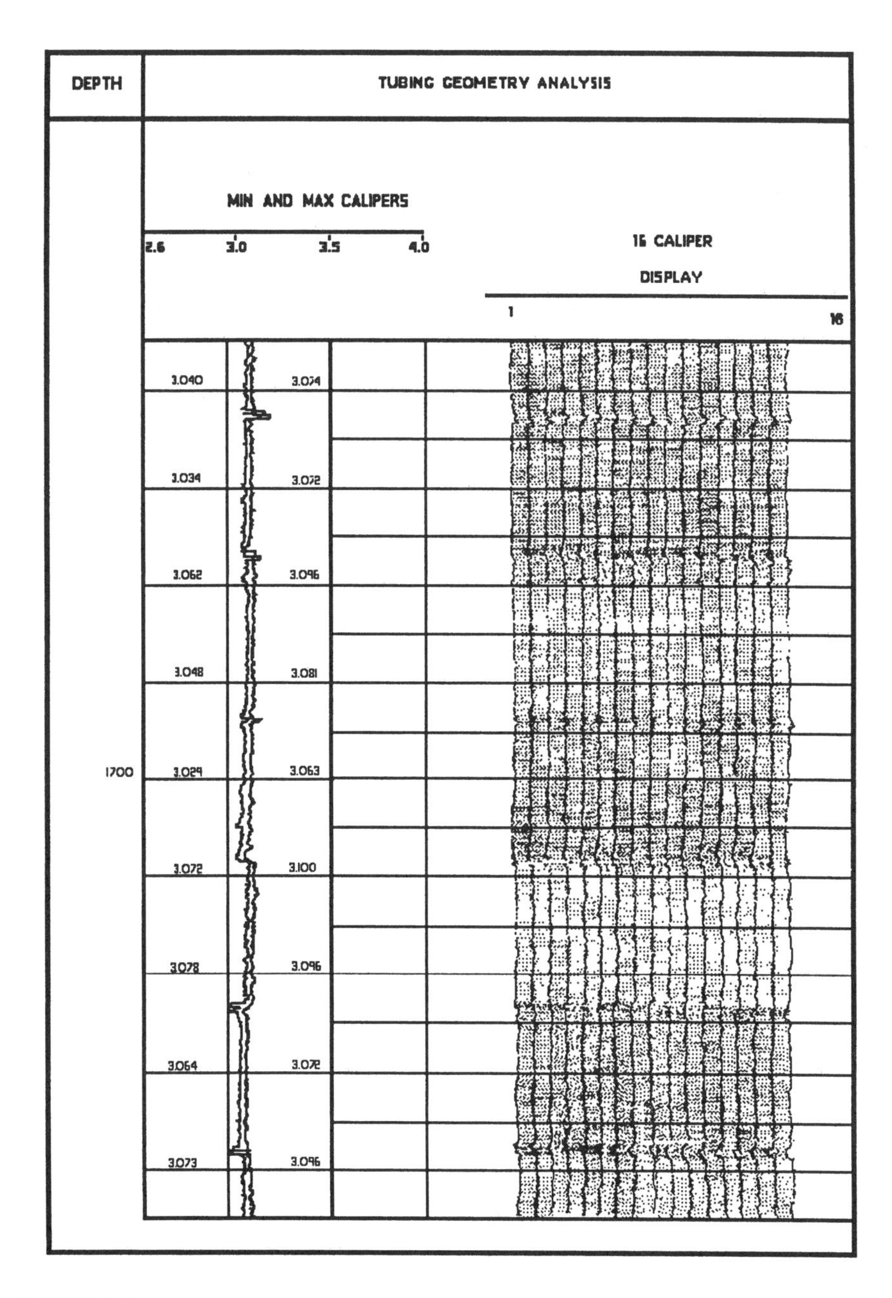

FIGURE 7.4
This is a log made in casing with a 16-arm caliper. (Courtesy of Schlumberger Well Services, Inc.)

travel time is measured, corrected for fluid character, and interpreted as hole diameter. This may be displayed graphically and/or numerically.

Acoustic calipers are accurate and do not damage the hole. In addition, they may be plotted on an expanded scale to locate possible casing damage,

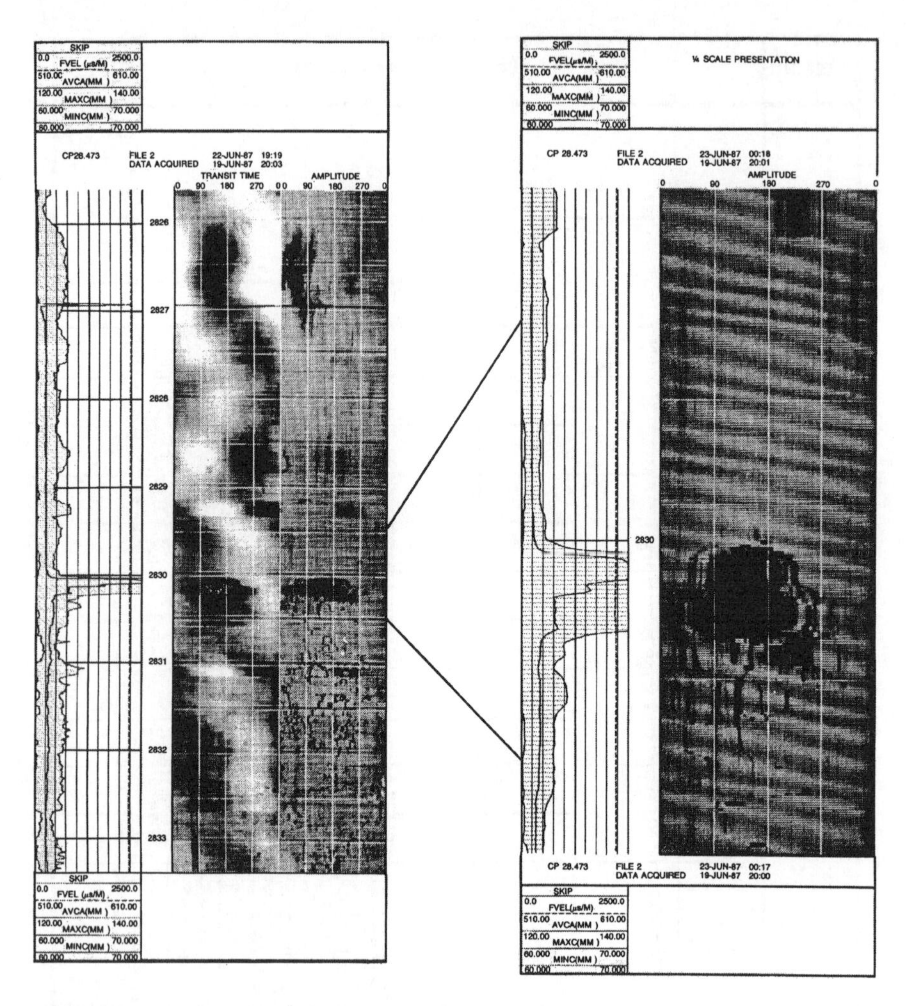

FIGURE 7.5
The BHTV standard presentation on left and expanded depth scale through the zone of interest on the right. (Adapted from Schlumberger Well Services, Inc.)

such as collapsed casing, damaged joints, splits, or corrosion. Figure 7.7 shows a typical casing corrosion log.

A modification of this principle used in the televiewer uses eight fixed transceiver transducers on a plane normal to the borehole axis. Schlumberger's unit is called the cement evaluation tool (CET). These are arranged 45 degrees apart around the tool. This system monitors the casing response to the high frequency bursts of mechanical (acoustical) energy. The degree of bonding to the annular cement is determined by analyzing the reflected pulses. The internal diameter of the casing is determined from the corrected transit time. The casing thickness is derived from the resonant frequency of the casing. Figure 7.8 shows a log made with the CET.

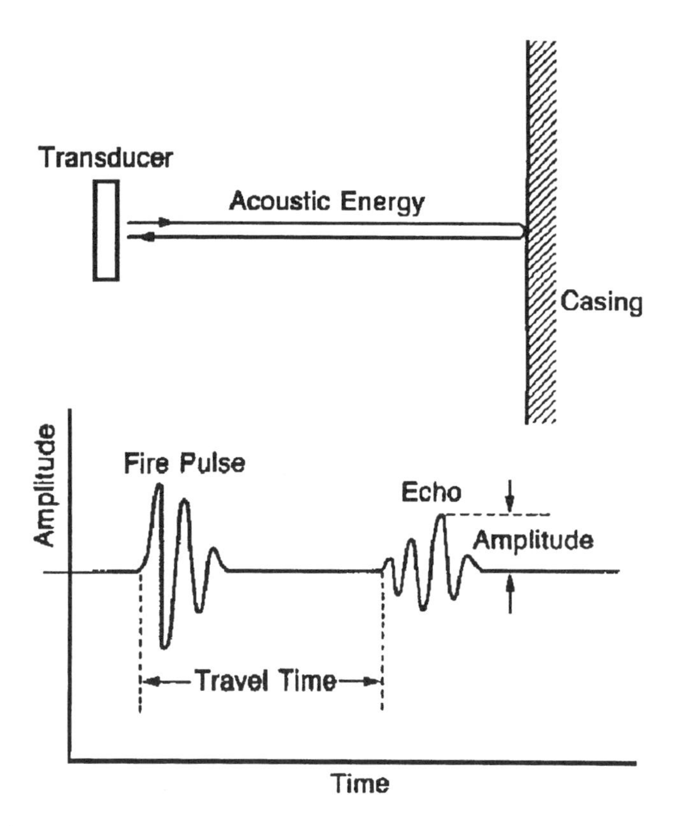

FIGURE 7.6
The principle of the BHTV measurement. (Courtesy of Schlumberger Well Services, Inc.)

7.7 Uses of Borehole Calipers

The caliper logs have many uses. The determination of hole diameter and volume was the original use and for many years, the only use. After a hole is drilled, it is often cased. Calipers are used to make sure that the hole is free of obstructions and is of sufficient diameter at all depths to allow the casing to be inserted freely.

The recording on the log is in units of hole diameter. Thus, with a three-arm caliper, the average volume, V_{ave}, of the borehole is

$$V_{ave} = \pi \left(\frac{d}{2} \right)^2 \Delta h \tag{7.2}$$

where

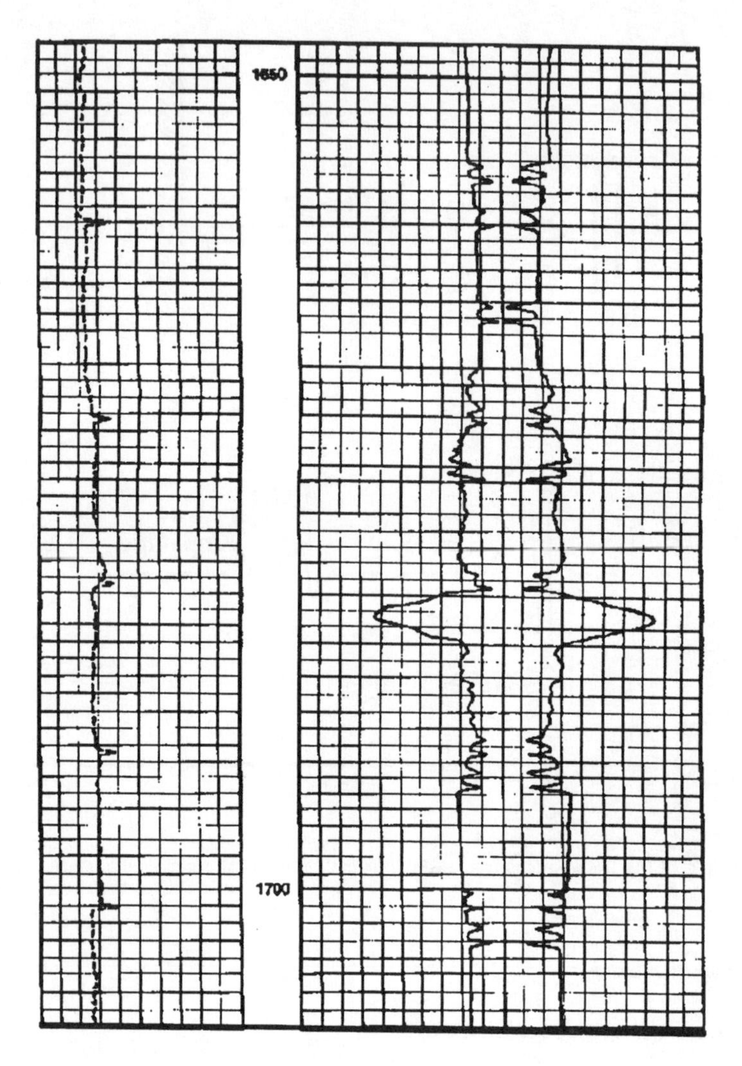

FIGURE 7.7
A profile log made in casing, showing severe metal loss. (Courtesy of Schlumberger Well Services, Inc.)

d = the average caliper reading.
Δh = the borehole depth interval.

A closer value of the borehole volume can be found by using

$$d_t = \sum_{h=1}^{n} \Delta d_h \tag{7.3}$$

where

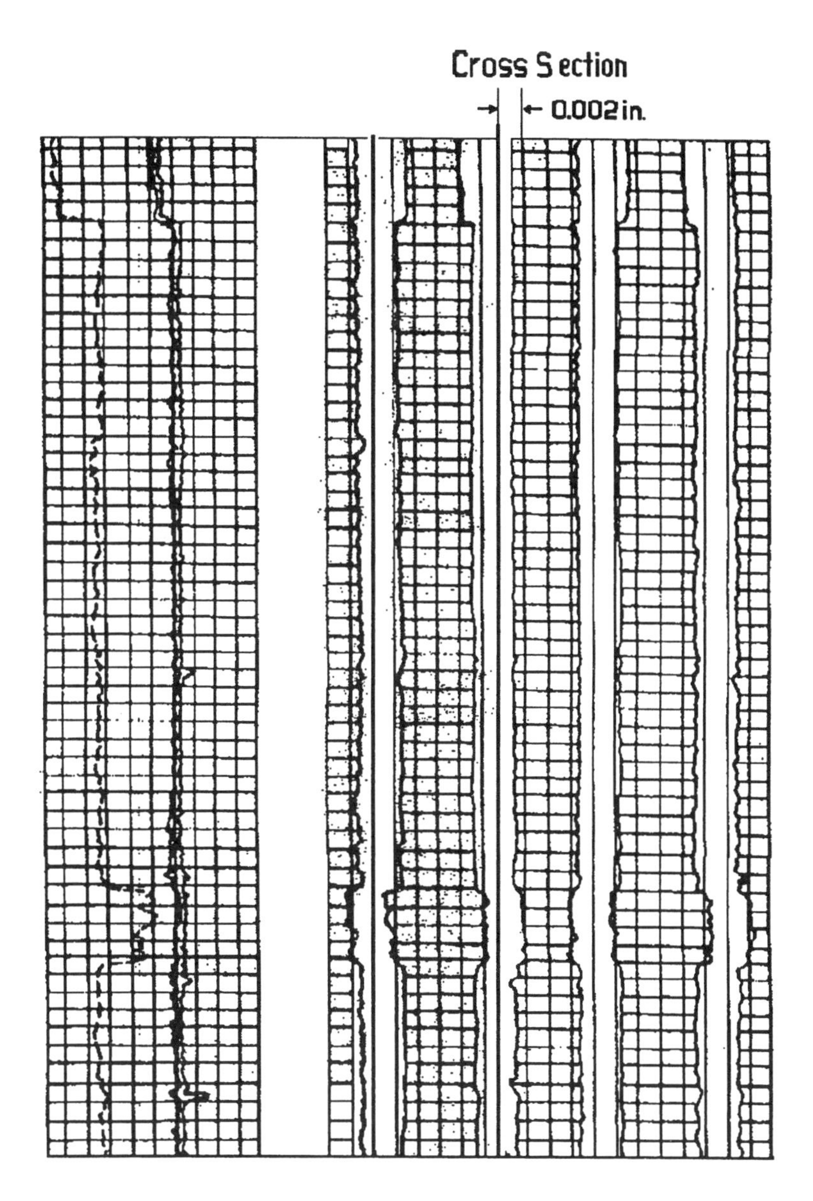

FIGURE 7.8
A CET log. (Courtesy of Schlumberger Well Services, Inc.)

d_t = a representative mean of the caliper deflection.
Δd = as small a depth interval, h, as practicable.

This averaging of the caliper reading can frequently be done by the on-truck computer.

Further, the volume of the annulus between the casing and the hole wall is frequently filled with cement and other materials. Knowing the volume of this space is a logical economic step. This volume is simply the borehole volume minus the outside volume of the hole casing. Determining the shape of the hole cross-section was the reason why the earlier three-arm calipers were replaced by multiple, independent measurements. The more accurate determination of the annular volume allowed savings in cement and other fill material costs.

When sidewall devices such as sidewall resistivity, density, and acoustic systems are operated, the contribution to the signal by the mudcake on the wall of the hole must be removed from the signal. In this case, the actual hole diameter is not as important as the thickness and character of the mudcake. The character of the mudcake (resistivity, density, water content, salinity, etc.) can be determined on the surface. The geometry of the mudcake within the measured volume is known. Thus, the influence of the mudcake character upon the measurement can be determined and the signal corrected for that factor. The reason for this is that the mudcake forms a substantial part of the measured volume of these tools. Because we are interested in the formation, not the character of the mudcake, its influence must be eliminated from the measurement.

The influence of the resistivity of the mudcake upon the sidewall resistivity devices probably requires the most careful correction. The range of mudcake resistivity can be from about 0.02 to over 5 Ωm, a range of 250%. Furthermore, the mudcake occurs at frequently, the most sensitive part of the measured volume. Therefore, detailed corrections must be made.

The influence upon the density system is not as great, in spite of the extremely sensitive volume of measurement in which it always occurs. This is because the density range of the mudcake is fairly small, from about 1.1 to about 1.5 g/cm^3, a range of only 31%. The most critical situation occurs when weighting materials (i.e., barite or hematite) exist in the borehole mud. The situation, as far as range is concerned, is similar for the sidewall neutron and the sidewall acoustic systems. Therefore, although corrections must be made, the amount of correction is small and not as critical as with the resistivity systems.

The caliper can be used in an open hole to determine something about the lithology of the formations. In a hole that is in good condition, the permeable sands will usually have a borehole that is the same diameter as the drill bit. A mudcake will form, however, because of the invasion of the drilling mud. Therefore, the sands will show hole caliper values that are smaller than bit size. This, of course, is the premise upon which the mudcake thickness measurement is based. Further, the tools that are used for this determination are designed to ride on the surface of the mudcake and not dig into it.

A rough measure of the mud quality can be determined by the mudcake thickness value. Good quality, low waterloss muds will have thin mudcakes. Salt muds also frequently have very thin mudcakes. These can sometimes be as thin as a millimeter. On the other hand, poor drilling muds, usually made with the natural, *in situ* clays, can have high waterloss values and thick, watery mudcakes. Invasions, with these poor muds, can be on the order of much more than a meter.

Competent shales frequently and massive carbonates always will have no mudcake. The wall of this type is usually smooth. Thus, the caliper will show a smooth, straight line at a value very near bit size. In hard rock environments, the caliper will frequently show fault locations. These are usually confirmed by indications on the microresistivity and density curves.

If a shale is highly laminar, however, it frequently will slough and cave to a diameter greater than bit size. Further, the wall will be rough. Thus, the caliper will show a rough, ragged line, with a value greater than bit diameter. The signature of a carbonaceous shale above a coal seam is characteristic. It frequently shows a graded cave immediately above the coal body. The floor shale seldom shows much evidence of laminations or caving.

8

Surface/Laboratory Methods

8.1 Introduction

Resistivity measurements are routinely made in the laboratory. In the oil and mineral businesses, these measurements are usually made on core samples. In engineering, geology, environmental studies, and real estate, the samples may be cores or they may be other shapes and sizes and from other sources than boreholes. These measurements often do not have the same weight that other measurements have, in the laboratory or in the field, in making decisions. Most core laboratories and many university geology laboratories are equipped to perform them.

Laboratory sample resistivity measurements have some good uses that should be carefully considered. They have some drawbacks, too, and particular problems. One important, and frequently neglected use, is as a teaching tool. The experiment in a freshman physics laboratory can show the concepts of resistance and resistivity as no lecture can. These measurements can also serve as a good research tool.

Laboratory resistivity measurements have the tremendous advantage over field measurements in that the laboratory has complete control over all of the variables, especially sample dimensions. Cross section, length, volume, density, and temperature are usually controllable or can be measured to any degree of accuracy desired. This is seldom true downhole and often not even on the surface in the field.

It is usually feasible to remove the contained (native) water from the sample and replace it with one of known composition and salinity. This is not always done properly, however. The native water must be removed without leaving a residue and without altering the physical and chemical composition of the solids in the sample. The included shale is especially susceptible to this type of alteration. It is also important that the geometry of the nondissolved salts not be changed by the exchange of the pore water. If the various precautions are observed, it becomes possible to determine a number of important parameters.

8.2 Samples

Samples can come from cores, outcrops, field rocks, cut rocks, or archived collections. Care must be taken to ensure that the sample is competent mechanically and free of any *unusual* inclusions, pores, and fractures.

The native fluid in any of the samples may be dried or partially dried. If possible, a sample of the native fluid should be collected with the sample. This last, of course, is almost impossible with outcrop and archive samples. It may be possible and feasible to transport the samples and to store them in their native fluid in plastic bags. This fluid should be analyzed for composition. If the sample has dried, some of the wash water will have enough of the salts to identify them. The native fluid of cores has probably been partially or wholly displaced by the drilling fluid filtrate. In any event, the fluid in a wet sample will, almost certainly, be a filtrate (that is, without solids).

Osmosis is certainly a possible error producer. If the laboratory fluid is more saline than the shale, it will tend to dehydrate the shale. If it is appreciably less saline, it will hydrate it. Thus, the foreign fluid must be as near the salinity of the shale as possible and the measurement should also be made as soon after impregnation as possible.

Another possible source of error is from the partial solution of and soluble crystalline material within the sample. This may be a salt, such as sodium or potassium chloride, or any number of common evaporites. This, of course, will result in a change of the pore geometry, with a consequent error in both porosity and permeability determinations. Again, the solution is to make the injected fluid as close to the native fluid as possible.

8.2.1 Porosity

Interconnected porosity is one major factor determining the resistivity of sediments. It is one of the reasons for running resistivity measurements. Of course there are many ways of measuring interconnected porosity in the laboratory: gas expansion, density measurements, acoustic travel time, etc. Resistivity measurements are simple, however, and can be quite precise.

The sample must be cleaned and dried without altering the composition and geometries of the solids within the sample. Then it must be completely saturated with a water solution whose composition and salinity are known. The salinity, if the sample contains any clay or shale, must be as close to that of the native water as possible. Electrode material is not greatly important, as alternating current should be used. Contact materials, such as electrode gel, porous paper, and cloth are available and should be used. Two-, three-, or four-electrode systems may be used.

The sample should be dried with warm, dry air. If possible, a lowered pressure atmosphere helps. The saturating water should be as near the salinity of the native water as possible to prevent osmosis from altering the dispersed shales. Vacuum and pressure impregnation will help overcome the surface tension of the water. The forces of this can be quite strong.

There is an advantage to using an electrode material that has the largest surface area possible. This will reduce the current density at the surface and minimize the electrode surface effects. It is a good idea to stay away from aluminum and stainless steel, as these form highly resistive and tough coatings which can greatly alter the electrode surface character. Iron makes a fair material, as it rusts and forms a large surface area. Lead is satisfactory as its surface coatings are usually nonrectifying. Copper, brass, and bronze can form rectifying coatings which can cause errors. Metal-metal chloride electrodes are excellent, but difficult to use and often expensive. Contact should be made through a gel or a thick, porous, saturated piece of paper between the electrode and the sample.

Either two-, three-, or four-electrode measurements can be made. The two extreme electrodes should be used for the current. The frequency should be kept low; 60 Hz is a convenient frequency. In special cases or for field instrument design, a lower frequency should be tried. Many surface resistivity/induced polarization (R/IP) units use frequencies below 10 Hz. Frequencies above 1000 Hz are used in some commercial instruments, but there is danger of errors, due to reactive effects, especially in some shales.

8.2.1.1 Interconnecting Porosity

The interconnecting porosity or effective porosity (ϕ_e) can be used to check on the validity of downhole measurements, especially the choice of the cementation exponent, "m". It can be used to determine the amount of shale in a sand and the presence of isolated porosity. Once ϕ_e is known, permeability, volume, and surface tension and capillary pressure determinations can be made with more confidence.

Porosity and permeability determinations, or even estimates, furnish valuable clues for subsurface fluid flow. This becomes important in water wells, particularly in high volume commercial wells, *in situ* leach mining, dam and wall footings, pond and lake construction, and waste disposal projects. Resistivity measurements, backed up by laboratory verifications, make an economical way to handle some of these problems. Both surface and downhole measurements are valuable.

8.2.1.2 Isolated Porosity

The isolated porosity (ϕ_i) is that which is isolated within the rock and has no communication with ϕ_e and is ignored by electrical methods. This value can be determined by measuring the porosity by density and/or acoustic methods. Both of these methods measure total porosity. Thus, if ϕ_e is subtracted from the value of the total porosity, the difference is ϕ_i. This value becomes important in the use of stone as a building material. It is valuable whenever rock is a load-bearing material.

8.3 Shale Determinations

Laboratory determinations can also furnish valuable data about resistivity of the shale in a shaly sand or carbonate:

1. A gas porosity measurement should be made on the dry sample. That is, there should be no free water present. The drying should not dehydrate the shale.
2. The sample should then be vacuum impregnated with a solution of sodium chloride (NaCl). This solution should have a resistivity or salinity as near to that of the shale as possible to estimate. This precaution is to reduce shale changes due to osmosis.
3. The resistivity and temperature of the sample should then be carefully measured.
4. Correct the bulk resistivity and the water resistivity for any temperature changes from the initial conditions.
5. Treat the shale resistivity and the known solution resistivity as parallel resistors resulting in the bulk resistivity. Solve for the shale resistivity.
6. Errors will be caused by the shale dehydration during drying, temperature changes, and osmosis with the new solution.

8.4 Density

Laboratory density measurements can be deceptively difficult to perform accurately. Most often, problems arise because of surface tension effects. Among other things, a small, entrapped quantity of air can cause substantial errors. Small samples must be carefully prepared for density analysis. If possible, a small sample should be cut to a simple geometric shape — a rectangular cube or a sphere. Gross dimensional measurements can then be made quite accurately with a caliper.

If a simple geometrical shape is not attainable, the volume of the sample can be measured by displacement. In the case of a porous sample, the gross volume can be simply measured by immersing the sample in a liquid that will not wet nor invade the sample. Liquid mercury is commonly used. If the sample is not porous, water or oil can be used. Mercury can be dangerous, however, if not properly handled. Care must be taken to avoid breathing the vapors and to avoid ingesting the liquid, even in minute amounts. Present laws also demand certain procedures for the disposal of mercury or mercury-contaminated equipment.

Measuring the volume of the rock matrix of a porous sample is difficult because the surface tension of the immersion liquid tends to prevent the liquid from completely penetrating the pore space. If the porosity of the sample is known or if it can be measured, the pore volume can be subtracted from the displacement volume to obtain the matrix volume. A small amount of detergent in the water will sometimes help. It is sometimes feasible to lower the pressure above the liquid until the sample is totally immersed. Then the pressure is returned to normal or increased slightly. This method has the disadvantage that the boiling temperature of the liquid is lowered and sufficient amounts can evaporate to cause errors. The operation must be performed rapidly and with a minimum lowering of pressure. The increase of the total volume is the sample volume.

The sample may also be placed in a chamber of known volume and the pressure lowered to 0.5 atm. The volume of air extracted will be equal to the volume of the chamber minus the matrix volume of the sample. This procedure can also be performed by doubling the chamber pressure. This method is also good to determine the pore volume of the sample.

Once the volume of the sample or the matrix of the sample is known, it can be dried and weighed. The sample must be carefully dried before weighing. Many types of rock may simply be oven dried, at a moderate temperature, for 24 hr before weighing. If this method of drying is used, however, the sample must contain no organic, volatile, or carbonate material. This includes mercury compounds and coals, as well as limestone and dolomite. Carbonates can lose their water or crystallization at surprisingly low temperatures. This author has noted such loses on samples of limestone dried for 24 hr at 350°F (177°C). Argillaceous samples can also lose substantial amounts of water during drying. They can also absorb water.

The density is simply the weight divided by the volume. In geophysics, it is customary to use the metric figure grams per cubic centimeter (g/cm^3). Seismic literature, especially older papers, often used pounds per cubic foot.

8.5 Spectroscopy

Laboratory spectrographic analysis can take several forms, each having advantages and shortcomings. Field results should always be confirmed by laboratory analysis. When this is done, confirming results can increase the level of confidence in the overall results. This can also provide data for correcting and normalizing field data.

8.5.1 Emission and Absorption Spectral Analysis

Emission spectroscopy can be applied to solid samples, liquids, or gases. The sample must be "excited" by energy: heat, light, or other forms of electromagnetic energy. If enough energy is supplied to a sample to exceed the *threshold* energy, one or more orbital electrons will absorb enough to be removed from their normal orbits (excitation). When the electron returns to its normal state, the excess energy will be emitted at a wavelength characteristic of the atom from which it was removed. The threshold energy needed is

$$E_0 = h\nu_0 \tag{8.1}$$

where

h = Planck's Constant.
ν_0 = the threshold wavelength.

Absorption analysis uses a white light source. The light is passed through the sample. Characteristic absorption lines will appear in the spectrum. Either emission or absorption spectrometry can be effectively used. Emission methods are especially used for opaque solids. Absorption spectral methods are used for gases, vapors, and transparent liquids. The energy source may be a gas flame, an electrical current, an electrical arc, infrared or ultraviolet (UV) light, or high energy radiation, such as X-rays, gamma rays, neutrons, or beta or alpha particles. Elemental and/or molecular composition can be determined, depending upon the instrument, the technique, and the sample type. Figure 8.1 shows the principle of the visual, emission, or the absorption spectroscope. The spectrograph is merely a recording modification of this instrument, usually incorporating a photosensitive device. Figure 8.2 shows an absorption spectrograph recording of an organic sample.

8.5.2 Mass Spectrograph

The mass spectrograph is used to determine molecule and/or atomic weights of gaseous samples. Usually, a gas sample is bombarded with electrons from a (usually) hot filament to convert the molecules or atoms to positive ions. The ions are then accelerated through an electric field and diverted through a magnetic field that is normal to the path of the ions. The amount of diversion of each ion will be a function of its velocity and its mass. Thus, the ions can be separated by atomic weight. The mass spectrograph is also used to separate isotopes, such as separating ^{235}U from ^{238}U. A portable mass spectrometer was developed for field analysis use from an automobile by Dr. J.C.M. Brentano of Northwestern University and Cambridge University.

The mass spectrograph ions are accelerated by a voltage field, V, collimated, and then focused into a normal magnetic field. The ion energy, E, is

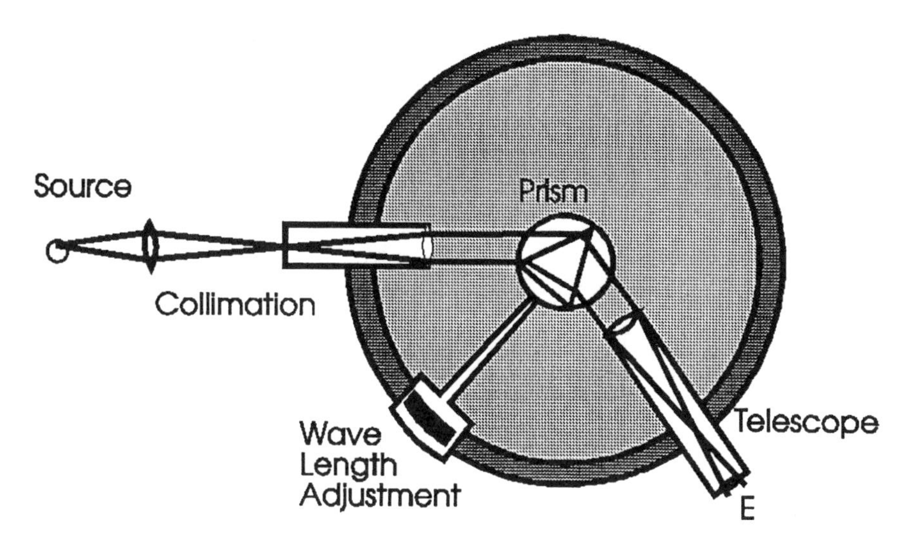

FIGURE 8.1
The principle of an optical spectroscope.

$$E = qV \tag{8.2}$$

where

q = the charge, in coulombs.
V = the accelerating voltage, in volts.

The kinetic energy of the ion is

$$qV = \frac{1}{2}mv^2 \tag{8.3}$$

where

m = the mass of the particle.
v = its velocity.

or,

$$v^2 = \frac{2qV}{m} \tag{8.4}$$

The force, F, of the magnetic field is the product of the field strength, H, the charge on the particle, q, and the particle velocity, v:

$$F = Hqv \tag{8.5}$$

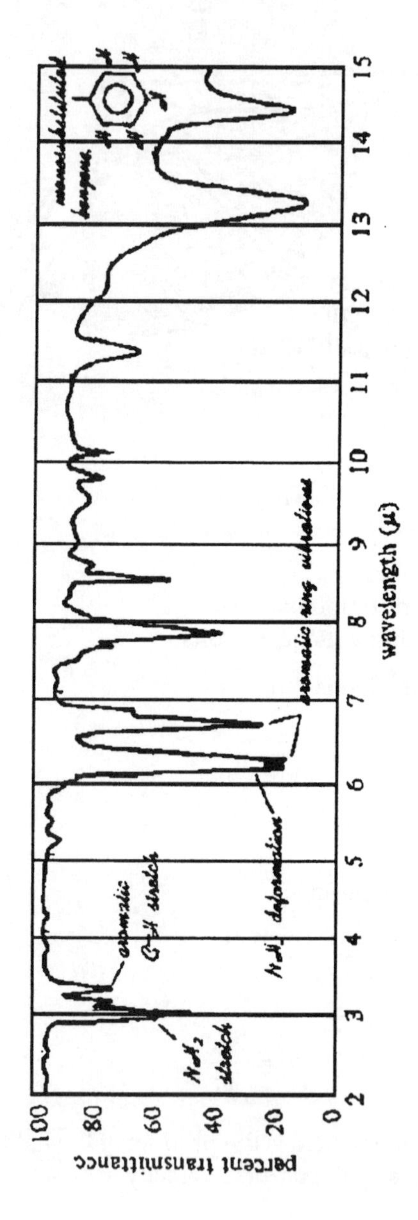

FIGURE 8.2
An absorption spectrograph. (Courtesy of Macmillan Publishing Co.)

The force on any particle in a circular arc, is given by mv^2/r, where r is the radius of the arc:

$$F = Hqv = \frac{mv^2}{r} \tag{8.6}$$

The charge to mass ratio, then, is

$$\frac{q}{m} = \frac{2V}{H^2 r^2} \tag{8.7}$$

Figure 8.3 shows the principle of the mass spectrograph.

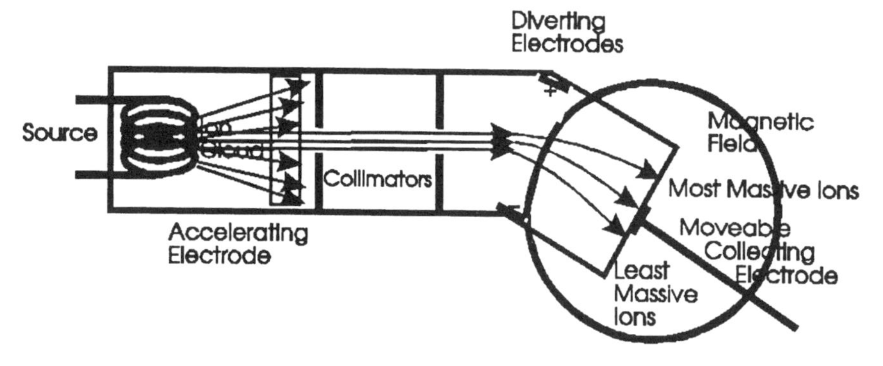

FIGURE 8.3
The principle of the mass spectrograph.

8.5.3 Spectral Radiation Analysis

The spectral analysis of natural radiation, such as gamma radiation, beta particle emission, or alpha particle emission can be performed in the laboratory or in the field. It is routinely performed on the surface and in air, as well. The advantage lies with the laboratory methods, compared to field methods and especially to downhole methods. This is because of the size and geometrical constraints on downhole systems and the weight limitations of surface field equipment. There is a further limit, also, to downhole systems, because they are subject to high ambient temperatures. The high downhole temperatures limit the performance of circuit components and cause high background noise in detectors. In the laboratory, the detectors may be large, have an optimum geometry, and be operated at low and/or constant temperatures.

The spectral analysis of photons and other particles depends upon absorbing all of the energy of the particles in the sensitive volume of the detector. This is discussed at length in Chapter 10 of this book and will not be covered here. The total energy of the particle must be dissipated within the detector in order to have a valid measurement of its energy.

All of the remarks concerning natural radiation analysis apply to activation analysis systems, as well. The bulk of the instrumentation for neutron activation analysis is almost identical to that designed for natural gamma spectral systems.

8.5.4 Laboratory Activation Analysis

Laboratory activation analysis follows much the same pattern as the downhole activation analysis described in the *Introduction to Geophysical Formation Evaluation*. There are some differences, however, most of which turn out to be advantages for the laboratory methods.

Surface equipment has a big advantage over downhole equipment. Borehole equipment must fit into the borehole. In the laboratory there are few such restraints. Thus, larger sources or electronic sources may be used in conjunction with storage facilities, plenty of power, cryogenic detectors, and multichannel analyzers. All of these things are limited in downhole use and even surface field use.

Neutron activation analysis, in the laboratory, can work with extremely small samples. Small sample results can also be misleading, however, as the sample is not statistically representative. The sensitivity of the systems can more than make up for this problem. Activation analysis logs should *always* be backed up with laboratory analyses on samples from cores and cuttings.

8.6 Cores

Coring and core analysis are popular in nonhydrocarbon exploration because so much of the exploration and development is done in hard rock environments. This means that a carbide or a diamond bit must be used, and it is almost as easy and economical to use a coring bit as it is to plug drill, under those circumstances. Therefore, a large percentage of mineral exploration depends upon core analysis.

Core analysis determines most of the parameters that wireline geophysical logging and surface measurements can. Most core laboratories are set up to measure porosity, density, and electrical resistivity. Some can measure acoustic travel time and some can do neutron activated and natural gamma ray spectral analysis. Most mineral and metallurgical laboratories can do X-ray analysis. Some have electron microscopes that can do an X-ray analysis on a specific spot or an impurity. Of course, cores and samples are necessary for chemical analyses.

One should keep in mind, though, that the sample size from a core is small compared to the sample size of most geophysical measurements. This means that the statistical variation (probable error) will be quite large compared to that of the smallest measured volume of surface or downhole geophysical measurements.

Assume that a chemical or X-ray analysis uses a sample size of 2 cm^3 (0.12 $in.^3$). A mineral-type density log examines a sample of about 10 in. (25 cm) diameter, a volume of about 8200 cm^3 (about 500 $in.^3$). A surface resistivity will measure a volume of about 1.5×10^{10} cm^3 (3.4×10^9 $in.^3$). Thus, for average values, the probable error for the core sample may be as much as 5.5×10^{19} times greater than the surface measurement and about 1.7×10^7 larger than the density reading. On the other hand, the detail of the core sample is about 4000 times better than the density and about 7.4×10^9 times better than the surface measurement.

This crude comparison means that one cannot depend exclusively upon core values to take the place of geophysical measurements. One must compare one with the other and use the combination that gives the best results.

Core samples are excellent as checks for logged porosity values. The corrected, logged porosity values are best for determining the volume of the reservoir. Core samples are indispensable for chemical and X-ray analysis. Borehole logged values are necessary for depth control of the cores. It is very easy for the driller to be in error by one or more joints of pipe. This can be a disaster in an *in situ* leach program. It can be very costly in sinking or excavating a mine.

8.6.1 Laboratory Resistivity Measurements

Many oilfield and mineral laboratories are set up to do electrical resistivity measurements. From these it may be possible to check logged values, calculate porosity, and estimate clay content. If the sample is competent, it is often cut to a convenient geometrical shape (i.e., a cylinder). If it is granular, it is usually packed into a plastic or glass cylinder. The length and cross section are carefully measured. In the case of a porous sample, it is often saturated with a known solution. A current electrode is then placed at each end of the sample. These may also be used as measure electrodes, but it is better to use two other electrodes, a distance L′ apart. A known current, I is flowed through the sample and the potential drop, E, is measured across the electrodes L′ or the current electrodes, L. The electrical resistivity is then calculated:

$$R = \frac{E}{I}\frac{A}{L} \tag{8.8}$$

Figure 8.4 shows a diagram illustrating the principle of the laboratory equipment for measuring core resistivities.

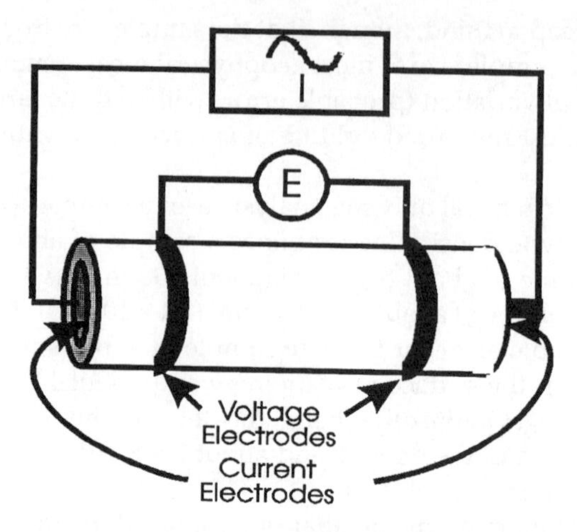

FIGURE 8.4
The principle of the laboratory equipment for core resistivity measurements.

8.6.2 Laboratory Acoustic Measurements

Oilfield core laboratories may also be able to make acoustic travel time measurements on core samples. From these, the porosity, rock type, and mechanical parameters of the sample may be calculated. Usually the sample is vacuum saturated with a known salt solution and enclosed in a void-free container. Transducers are then placed in contact with each end of the sample and a pulse is sent through it. The travel time is read and the received pulse is usually displayed on a calibrated oscillograph. The rock type can be determined by the travel time, t, of the pressure wave (P-wave). The porosity, ϕ, can be calculated from the travel time of the P-wave, if the rock type is identified:

$$\phi = \frac{t - t_{ma}}{t_f - t_{ma}} \tag{8.9}$$

where

t = the measured travel time.
t_f = the travel time of the pore fluid.
t_{ma} = the travel time of the matrix rock.

Mechanical parameters and other rock mechanics values can also be calculated.

8.6.3 Laboratory Density Measurements

Laboratory density measurements are usually done by the displacement method. If the sample is porous, it may be flushed and dried or it may be saturated with a known fluid. The sample is then carefully weighed on an accurate balance. It is then immersed fully in a noninvasive fluid and the volume carefully noted. The density, ρ, is simply

$$\rho = \frac{W}{V} \tag{8.10}$$

where

W = the weight of the sample.
V = the measured volume of the sample.

The porosity of the sample and its rock type can be determined. The resulting values can be used to determine premining parameters, clay content, and corrections for acoustic, seismic, and gravity measurements.

8.7 Coal Parameters

Coal development requires the knowledge of four parameters: heat, moisture, ash, and volatile contents. The moisture and volatile contents can be determined by fractional distillation. The sample is heated gently and the liquids are driven off as vapors. Upon condensation, the water and the organic-based liquids can be separated and recovered. The heat content is determined in a bomb calorimeter. The sample is heated to destruction and the heat of combustion is measured. The residue is the ash content. Sulfur content is usually measured by chemical analysis.

9

Uses and Analysis of Geophysical Methods

The petroleum industry has effectively combined surface and borehole geophysical methods. To some extent, the mineral industry is beginning to do the same. Airborne gamma ray and magnetics surveys and surface gamma ray measurements have been used in the exploration for heavy metals for many years. The uranium effort, during the period between 1950 and 1975, introduced borehole geophysics to mineral exploration and, later, to mineral development. These methods have been picked up by the coal industry and hydrology, to their benefit.

The many disciplines discussed in this text — engineering, real estate, highway construction, the building industry, waste disposal, flood control, geology, even the legal profession — should seriously consider the use and economy of geophysical methods and the expertise of the associated geophysicists and geologists.

9.1 Single Measurements

Early measurements were handled entirely by analog methods and were usually of a single type. The early gravity measurements covered wide areas. Measurements were made and the data reduced by purely analog methods. Eötvös' first exhaustive field investigation (1901) covered the entire Lake Balaton while it was frozen. This must have taken weeks to months to make the measurements alone. These were then assembled in an isogravity map. Kelvin recorded the temperature of a water well from top to bottom, point-by-point, as the first borehole log (late 19th or early 20th century). Even this simple measurement must have take hours to record.

It is seldom, however, that enough information can be gathered from a single curve of a single type of measurement. A resistivity curve, for example, seldom can point out the differences between a limestone, a dolomite, and a basalt, even though two are sediments and the last is igneous. Even the presence of petroleum or very low salinity water can be confusing if only one type of curve is available. It is like trying to identify colors and shades from a black-and-white photograph.

If a curve of another type of physical measurement is added, however, the available conclusions are more than doubled. If a third curve is added, the usable data increases almost exponentially. Notice in Figure 9.1 how uninformative one curve by itself is, but with four curves and core information one can form a complete idea of the zone make-up.

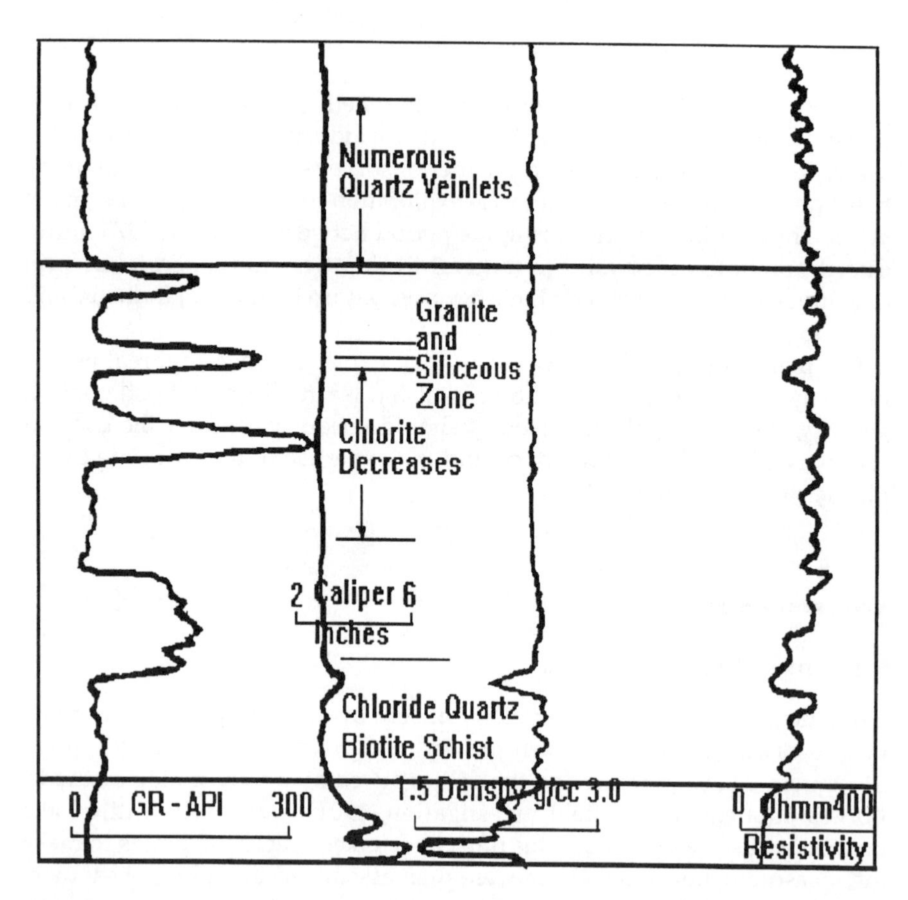

FIGURE 9.1
A hard rock mineral log section with four curves.

A type of measurement, however, must be chosen to fit the situation and with a goal in mind. A resistivity curve, for example, has only limited usefulness in a massive igneous bed, unless it was designed for that type of work or unless the bed is highly fractured. An oilfield density curve is useless in a massive basalt, because it was designed for lower densities. A spontaneous potential curve is useless in an air-drilled or oil-filled hole, because it measures the presence of ions. An uncalibrated oilfield gamma ray system is limited and swamped when looking for radioactive fuel minerals.

9.2 Development of Multiple Curve Measurements

Mechanical recording methods were a great time saver when they appeared, increasing accuracy. Early recorders in a logging truck used optical galvanometers, recording with a beam of light on photographic film. The film, in the case of the borehole logs, was driven by the logging cable. Thus, the recording was made as a function of borehole depth. More galvanometers were added as simultaneous measurements were added. Late recorders used nine galvanometers to record four simultaneous measurements. With this system, the information could be monitored during logging. It was available within minutes after logging was completed. A cursory examination was done in the field, after the film was developed. Detailed analysis of the logged information was done in detail in the office. This process took days to weeks. Airborne measurements of magnetic fields and gamma radiation were made in strips and assembled, later, into maps of isomagnetic and isoradiation lines. This process required weeks of time.

The trend, however, was to record more and different types of measurements. Instrument design also tended to emphasize a multiplicity of simultaneous measurements. One set of measurements by one single type of instrument is almost invariably limited and is seldom used.

From this beginning came the modern practice of recording almost any combination of simultaneous and sequential curves. Use of computer correlation methods, data reduction software, and routine analytical methods have made the process more accurate, faster, and more valuable.

The tremendous value of modern geophysical methods lies in its multiplicity of types of measurements. This, of course, results in a tremendous amount of information. One simple mineral log, with a minimum sample rate, collects 90 data points per foot (295 data points per meter) of hole for each curve. Most devices collect much more data than this. The advent of modern computer data handling methods is beginning to open an analytical vista only dreamed of twenty years ago.

The Schlumberger brothers realized early that one curve was not definitive enough. The first log made by them was a single resistivity curve. The spontaneous potential (SP) curve was added early. Resistivity curves of various depths of investigation were soon added.

Because there is a practical physical limit to the number of instruments that can be assembled for one set of measurements, multiple runs of borehole measurements use a common correlation curve. In an open hole, the usual curve is an SP curve. In a cased hole, the usual correlation curve is the gross count gamma ray curve. These two curves both respond to the shale content of the formation. Thus, they can be correlated with each other (see Figure 9.2). Notice that the SP curve, if it were reversed, would closely resemble the gamma ray curve and can easily be correlated. In this case, the SP curve is

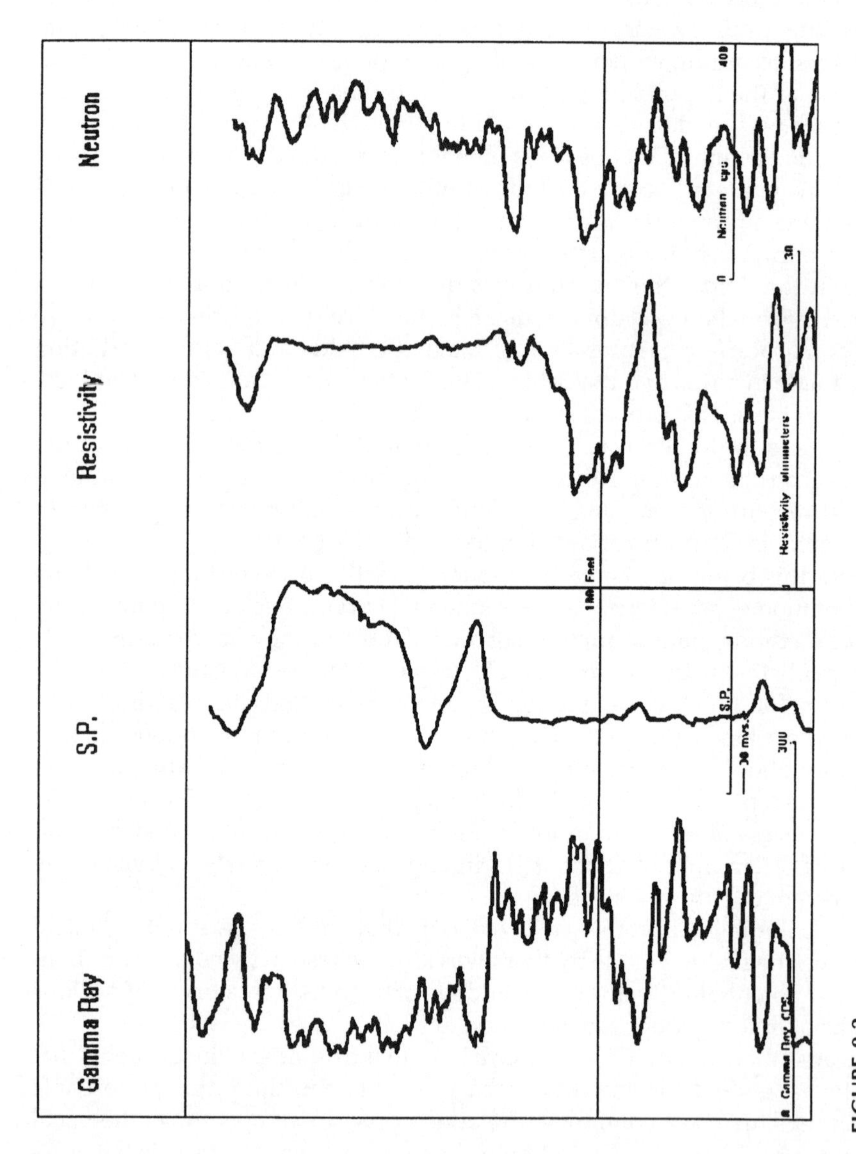

FIGURE 9.2
A water well log.

reversed because the drilling fluid had a lower salinity than the formation water.

9.3 Checks

As we record resistivity data (or any data, for that matter) we can make several checks on the validity of the values. Normally, we would expect these values to be "typical" for the circumstances. Of course, we do not know the exact values or there would be no point in making the measurement. If they do not fall within the typical or expected range, however, this fact may be important. Even if it is impossible or impractical to remake the measurement, the very fact that we know whether the measurement is valid or in error is valuable. If it is correct, it is an anomaly. If not, it indicated a malfunction. It is vital to decide if the measurement is in error or if the values are real, before leaving the measurement site. Figures 9.3 and 9.4 show typical mineral hard rock logs, compared with the core information and core depths.

There are many things that can affect our measurement and cause aberrant values. For example, we would normally expect a zone of massive granodiorite to have a resistivity above 10^8 Ωm in ambient temperatures below 300°F (150°C). Actually this would probably be beyond the range of the usual oilfield resistivity equipment. If the resistivity is lower than that, as suggested by normal readings elsewhere in the zone or borehole, the possibility that the rock is fractured or contains metallic material (i.e., pyrite) should be entertained. Either of these variants can be very significant in a mining or other engineering project, not only as a potential ore body, but, in the case of fractures, because of the potential danger during excavation or later structural failure. Figure 9.5 shows a mineral log that has a caved section of the borehole. This caving might not be evident without the caliper log, although the density curve should suggest a problem.

In the event that metallic material or microfracturing is suspected, the surface and airborne measurements should be inspected. Metallic deposits will probably show up on the airborne magnetic survey. They may even show on the airborne or surface gamma ray maps. Surface resistivity measurements may show either metallic bodies or fracturing (if the latter is extensive). Seismic surveys will show any anomalous features. Gravity would show massive metallic, water, or gas deposits or caves. Figure 9.6 shows a log through a near-surface zone which has extensive sulfide mineralization, probably in discrete layers or laminae.

In petroleum evaluation, the anomalous situation is more well known. Actual porosities can be measured in several ways. We can predict the resistivity of a porous zone quite accurately. If the readings are higher than expected, we suspect the presence of hydrocarbons or gas. This is well known

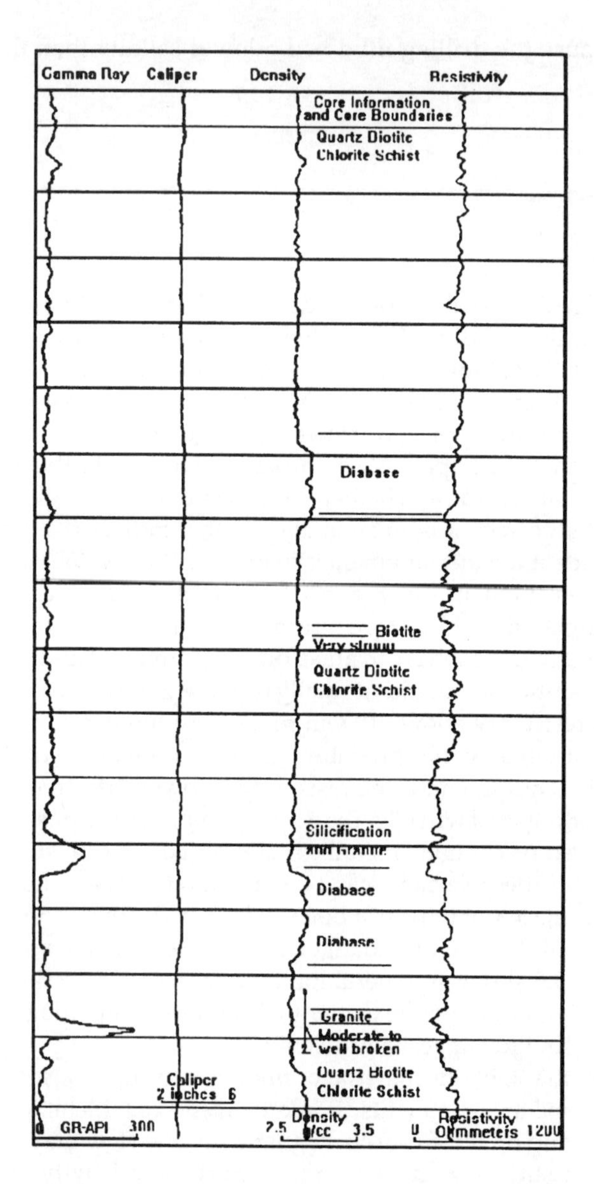

FIGURE 9.3
A mineral hard rock log.

(and covered in *Introduction to Geophysical Formation* and *Standard Methods of Geophysical Formation Evaluation*) and will not be covered here.

An anomalous low resistivity reading in a massive rock should be investigated, especially if mining is planned. There are several methods available. The microresistivity tools are quite sensitive to the presence of fractures. The microscanner resistivity systems can closely identify electrically conductive fractures in massive rock. The conductive fill material may be water or one of

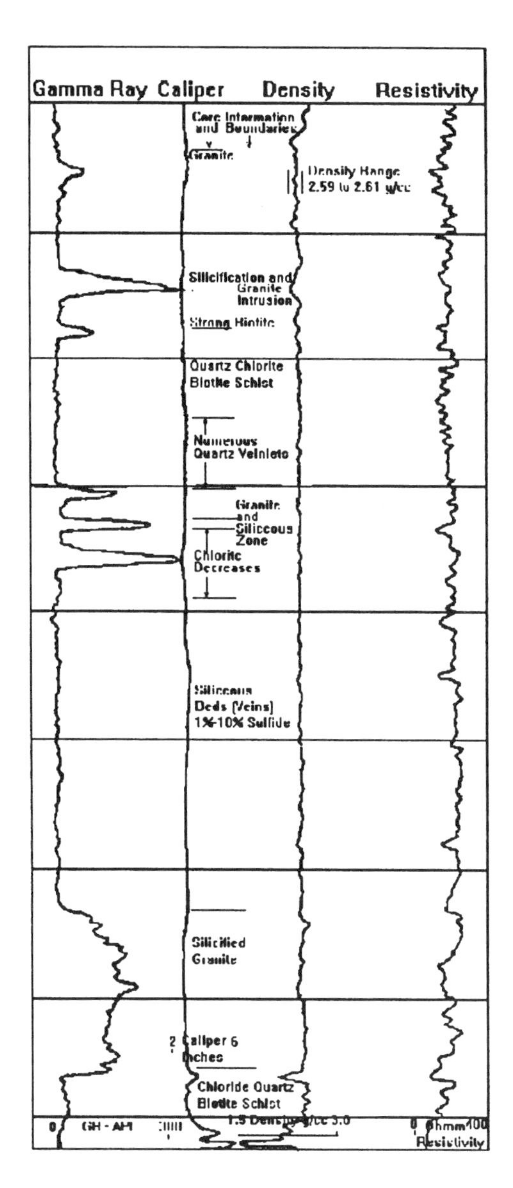

FIGURE 9.4
A mineral hard rock log through low sulfide concentrations.

the electrically conductive materials, such as shale, clay, sulfides, or native metal.

If the microscanner indicates conductive fractures, the next step is to identify the fill material. The natural gamma ray system will locate radioactive fill material. This, however, may be clay, sulfides, or deposits of uranium/thorium from water moving through the fractures. A neutron log can identify

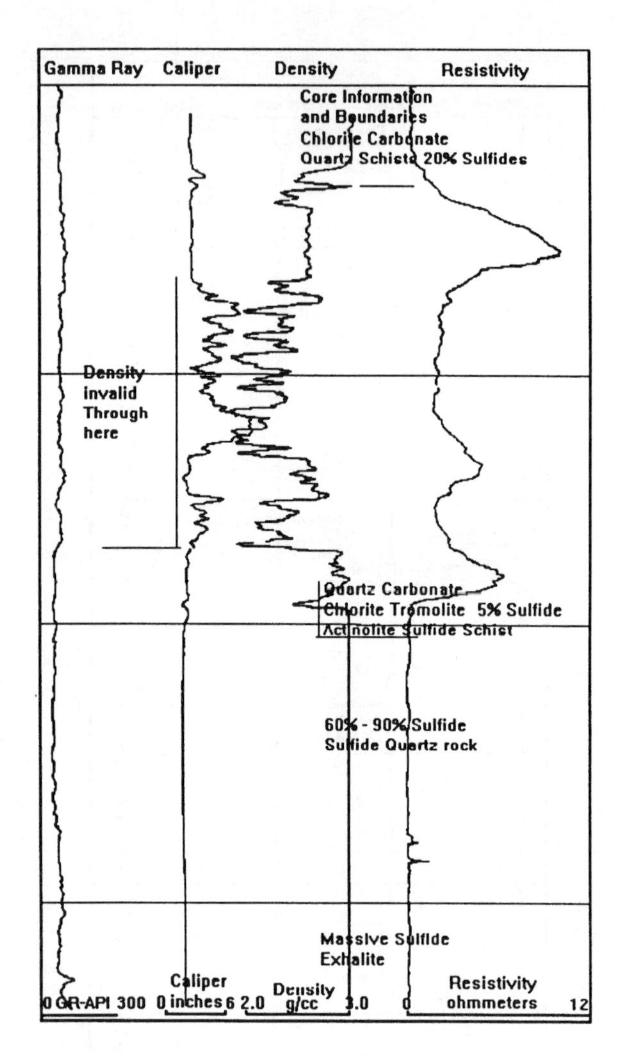

FIGURE 9.5
A mineral log showing a caved portion of hole and massive sulfides.

hydrogenous material (water and/or liquid hydrocarbon) in clays or open fractures. A density system can identify metallic fill.

9.3.1 Water Exploration and Evaluation

Underground water becomes a more and more valuable resource. As surface waters have tended to become more polluted and exploited, underground waters tend to be a desirable find. Underground water can be treated as a mineral when searching for it. Mineral exploration techniques can be successfully used.

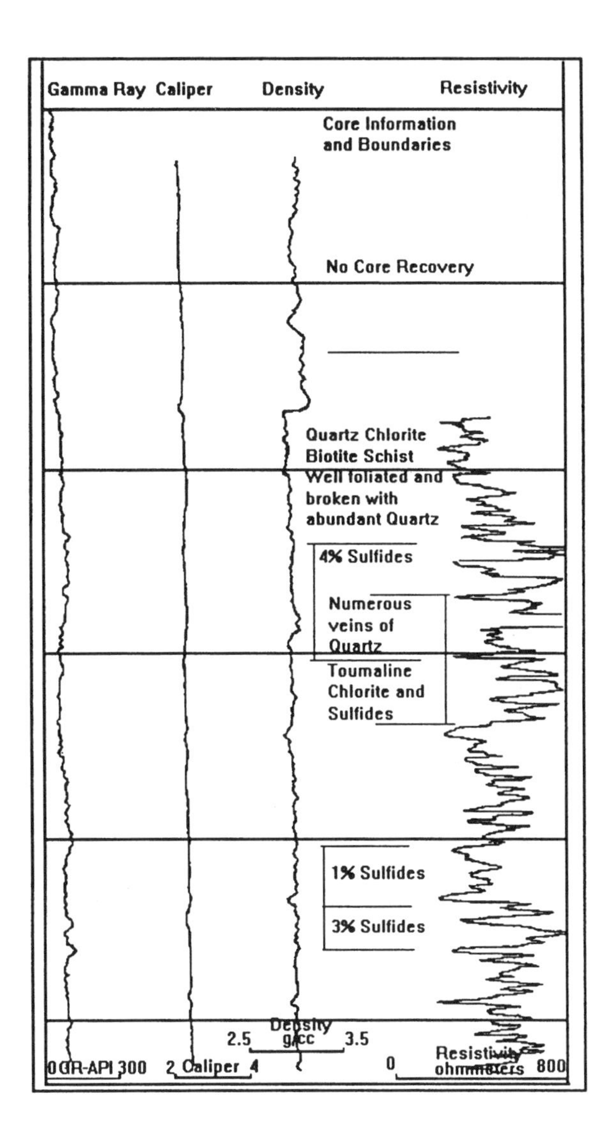

FIGURE 9.6
A mineral log showing a foliated, mineralized zone.

Water has a number of well known characteristics. Its density (ρ_w) will be about 1.00 to 1.22 g/cc. Its electrical resistivity will vary from about 50 to 100 ohmmeters to about 0.02 Ωm, depending upon the total dissolved solids (TDS) and temperature. It has a P-wave transit time (t_p) range of about 207 (679 μs/m) to about 170 μs/ft (558 μs/m). Its neutron capture cross section (b_w) range is from about 22.08 to about 146.2 b. Its thermal conductivity is from about 1.39 × 10^{-3} cal/cm sec/°C at 0°C to about 1.60 cal/cm sec/°C at 75°C. All of the foregoing depend upon the temperature and the dissolved solids content. Water is found filling the pore spaces of granular formations, fissures, fractures, and caves of carbonate zones. It may be found as a solid, but this is rare. Because of the pressures at which water is found, it is almost invariably found as a liquid. It is found at all presently practical drilling depths.

In addition to the obvious water needs for domestic uses, it is also needed for irrigation, commercial processing, and cooling or heating. The purity of the water will determine its suitability for any use or uses. Besides the purity, the producibility of the water is a vital factor in determining its use. Sometimes it can be used for multiple purposes, such as heating and household use or irrigation. Occasionally it is used as a vehicle for carrying other minerals. Therefore, a number of factors must be taken into account to determine the economic suitability of a water deposit.

The applications in a water well project, for example, could be pictured as running the gamma ray (GR), spontaneous potential (SP), and two or three resistivity curves in the open hole. This would locate the water sand, give an estimate of the shale content of the sand, and determine the depth of invasion of the drilling fluids. A caliper and SP (or GR) combination would then be run to determine the competency and the relative shaliness of the sand. The need for cementing or gravel packing will depend upon the purpose of the well. The locations of screens can be pinpointed accurately with respect to depth and/or with respect to the best producing zone of the aquifer. Their positioning can be verified afterward. Refer back to Figure 9.2, a water well log.

The actual amount of water in the sand (absolute porosity) can be determined with a neutron porosity, density, or acoustic porosity log. The absolute porosity must then be corrected for the presence of shale. This correction can be determined from the gamma ray log and/or the SP. A high percentage of shale will not only reduce the effective porosity, but it will also affect the permeability of the zone. If the relative amount of shale is high, the zone should be flow tested before casing. Please refer to *Introduction to Geophysical Formation Evaluation*.

After the hole is cased, the gamma ray and neutron-porosity logs can be run to be sure that a screen or the gravel pack was set in the proper place; or, in some cases, a perforating gun in combination with a gamma ray or a neutron log, could be run to produce the well.

The SP reading can be used to determine the eNaCl salinity of the water, before the well is cased. (Please refer to *Introduction to Geophysical Formation Evaluation*, Chapter 7.) It would probably be wise to take a water sample, especially if the salinity is high or near the acceptable borderline. This can often be done by the driller. Samples can also be taken with a fluid sampler. (Please refer to *Introduction to Geophysical Formation Evaluation*, Chapter 6.) The drawback to this last method, however, is that a large borehole is usually required. In the case of commercial water wells, this may not be a drawback. The advantage to taking a water sample is that it can be tested in the laboratory for dissolved solid type (ion type), actual salinity, and the acid-base state (pH). Some work has been done to try pH electrodes and specific ion electrodes downhole, but as far as the author knows, there is no such service offered.

The temperature of the well water can also be important. A high heat content, such as might be found in some zones or in deep holes, can be used for heating even if the water is to be used for drinking or irrigation. A temperature log can easily be run before casing. This has an added advantage of locating the water entry point if the hole is making water. This log should be run going into the hole, as the first of a suite of logs, so that an accurate reading is taken.

If the well is losing water, the SP curve can be used to pinpoint the water loss point. The SP system will detect the electrofiltration potential generated by the moving water. This should be run with and compared to the gamma ray curve. This can easily be done because the two curves resemble each other. Thus, the anomalous zone on the SP will be easier to spot.

9.3.2 Sand/Shale Sequences

The effectiveness of the combination of the SP curve and two (or more) resistivity curves to determine water saturation has been discussed for petroleum purposes. Please refer to *Introduction to Geophysical Formation Evaluation* and *Standard Methods of Geophysical Formation Evaluation*. This combination works equally well to locate the top of the ground water, the presence of perched water, and the thickness of the transition zone of a coal.

The high resolution resistivity devices can be used for estimating the probable existence of vertical permeability of a water sand when thin shale laminations are a problem. These are the focusing electrode devices and, especially, the sidewall focusing electrode tools. They are the laterologs and the microlaterologs and the related devices. The actual device should be chosen on the basis of the resolution desired. The device should be run in combination with a gamma ray and/or an SP log. The latter are to determine the actual amount of shale in layers.

9.3.3 Saturation

Introduction to Geophysical Formation Evaluation and *Standard Methods of Geophysical Formation Evaluation* discuss the effectiveness of using the SP curve and two or more resistivity curves for determining fluid migration and saturation (S_w) for petroleum purposes. This procedure and Archie's rules work equally well where gas or air occupies part of the pore volume. Thus, the determination of saturation can accurately tell the extent of the partially saturated zone above the top of the water table and gas-filled reservoirs. It may also be used, if care is taken, to estimate the possibility of draw-down in shallow or gas-topped deposits.

The probable shale content (V_{sh}) of the water-bearing zone can be found from the gamma ray curve, either the gross count curve or the thorium curve of the KUT system. This procedure is to be found in *Standard Methods of Geophysical Formation Evaluation*. This amount of shale must, of course be subtracted from the total porosity, if the porosity was determined from resistivity values.

The salinity of the formation water can next be found from the SP curve, in terms of eNaCl. If the component types of the dissolved solids are known, they can be converted into the probable amounts of the various components. These procedures (Dunlap Multipliers) can be found in *Standard Methods of Geophysical Formation Evaluation*.

A porosity log of some kind should be run to determine the total porosity of the producing zone. This can be either a compensated density (FDC), the Litho-density (LDT), or the neutron porosity system. Better accuracy and lithological information can be gained by cross plotting the density and the neutron porosity.

Knowing the values of S_w and using the resistivity and SP curves, one can now closely estimate the total reservoir volume, potability, and producibility for the zone of influence for that particular well. With the SP and resistivity curves, the best zone or zones to and where to set packers and screens or cement for greatest efficiency can easily be determined.

If the well is a large diameter commercial well, the nuclear magnetic resonance tools (NML) should be considered. Modern NML systems can tell the difference between the bound waters of clays and the free waters of the pore space. This has the added advantage of showing the effective porosity and the presence of any liquid hydrocarbons. The difference between the effective porosity, from the NML and the total porosity from the FDC or neutron porosity can be considered the clay content of the sand. The disadvantage is that the downhole tool has a large diameter.

9.4 Salinity Effects

Salinity can be described in may ways. One can state the amount of dissolved salt in the solution by weight, the amount by volume, its taste, its electrical resistivity or conductivity, the solution density, as equivalent sodium chloride (eNaCl), total dissolved solids (TDS), the ion concentration, and many other ways. It can be defined as the amount of salt dispersed as ions in a solution. The solution may contain a single salt type or many of differing amounts.

There are other things that enter into the picture, also. Salinity of the formation fluids is important. Temperature effects, of course, are part of the salinity picture.

All natural waters contain some amount of dissolved salts while some may contain very few. Areas in Brazil, for example, have public water supplies having 50 ppm or fewer of total dissolved solids (TDS). Casper, WY has public water with about 50 to 60 ppm of TDS. These low salinity waters can be very confusing in the petroleum industry, where geophysical exploration depends, in part, upon a high resistivity contrast between petroleum oils and formation waters.

The most common ions found in water are sodium (Na^+) and chloride (Cl^-). Potassium (K^+), calcium (Ca^{++}), sulfate ($SO_4^=$) and carbonate ($CO_3^=$), are fairly common. There are many others, but they are less common and usually of lower concentrations. Na^+ and Cl^- are so common that the mixed solutions are often described by equivalent sodium chloride by weight (eNaCl) concentrations. There are ways (i.e., Dunlap Multipliers) for converting from one to another method (see *Introduction to Geophysical Formation Evaluation*). Ions are formed when the dissolved salt breaks into ions; that is, when it dissociates.

Note that ions will have one or more unit electrical charges and will be either positive (+, metallic) or negative (-, nonmetallic). These are indicated by the superscript following the chemical symbol for the ion type, as Ca^{++}, for example. If it does not have an indicated charge it is presumably not ionized, or it is neutral. The positive ions, the cations, will be attracted to the negative-most part of an electric field (the cathode). The negative ions, the anions, will be attracted toward the more positive part of the electric field (the anode). They will also be diverted if they move across a magnetic field, the cations in one direction and the anions in the other. Their speed of movement, in either case, will be a function of the amount of charge, the size and mass of the ion, the viscosity of the fluid, and the intensity of the field. The force will be at 90 degrees to the magnetic field. Noncharged or neutral particles will be unaffected.

Ions may also be from materials other than salts, and the dissolved solids may not all be ionized (only partially dissociated). Ions may be from acid, bases, or gases. The nondissociated solids may be from weak acids or bases, from sugars, or other organic materials. In general, the nonionized solids do not have an effect upon the electrical resistivity of the material like the ionized ones. They do, however, affect some of the other parameters, such as density.

Formation water salinity, and thus, physical rock characteristics, will follow general patterns. A given rock type can be expected to have a normal range of salinity, density, travel time, and resistivity. Figure 9.7 shows the approximate resistivity/salinity relationships of some sediments. Some ion types and/or salinities are characteristic of certain formations and areas. For example, limestone areas generally have anomalous amounts of Ca^{++}, Mg^{++}, and CO_3^{++}. Marine deposits usually have high Cl^-, Na^+, and Mg^{++} concentrations. Trace amounts of metallic ions are often significant in metallic mineral exploration. Formation and dissolution of salt domes can be detected by the Cl^- and $SO_4^=$ concentrations down dip.

Ion content (salinity) is involved in many of the geological mechanisms and may be examined by geophysical methods. The success of the roll front

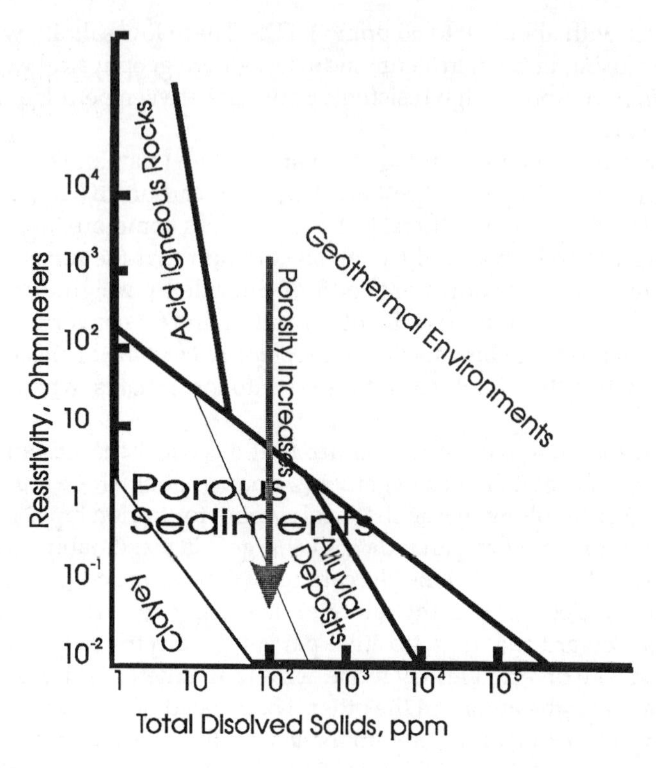

FIGURE 9.7
A chart of R vs. Salinity.

location by borehole geophysics is a good example of this. The location, mechanism, and analysis of shales and clays is another good example.

Resistivity methods are excellent for determining salinity. This is, of course, one of the reasons resistivity logging is a standard method for petroleum exploration. Unfortunately, several of the valuable companion logs are usually neglected when these methods are used for nonhydrocarbon purposes. Water well logging, for example, usually uses only the qualitative single point resistance log for stratigraphic purposes. For this it is excellent. However, a quantitative high resolution resistivity system, such as one of the focusing electrode resistivity logs (such as the three-electrode system) shows not only better bed resolution, but also allows a quantitative measurement of the formation water salinity. The SP and the gamma ray curves are equally valuable. If the borehole resistivity measurement is combined with the SP log, the formation water salinity can be separated from the invading fluid resistivities. Further, a gamma ray curve will allow a quantitative assessment of the clay content (and thus the permeability) of the zone. Also, if the borehole logs are combined with a surface resistivity or seismic survey, the extent of the water zone and its depths can then be measured. In addition, a water sample and a laboratory analysis can pinpoint the types of materials dissolved in the water.

Salinity is something that must be known about most water sources. The salinity determination value is not restricted to public drinking water. Cooling water must often adhere to the same or more severe restrictions as potable water. In addition, the temperature of that water can be determined from the borehole. This is also true of irrigation water and stock water. A little quantitative effort can save greatly during the development and use stages. A good geophysical analysis of an irrigation project can avert disasters, such as salinization of good crop land.

9.5 Oceanographic Uses

Electrical resistivity and conductivity (induction) methods can be, and are, used in oceanographic studies. They are used both in the water body and in the seabed boreholes. Surface array methods could be used, also, but as far as this author knows, this has not been done. Again, it is necessary to include a good temperature log with the resistivity or conductivity log.

It would be quite feasible, although this author does not know of such a use, to increase the sensitivity of the density logging system detection to observe the densities of the various layers in a deep-water oceanographic study.

9.5.1 Water Body

In the body of an ocean, lake, or river, the electrical resistivity (or its reciprocal, conductivity) are convenient and sensitive measures of the water salinity, either in terms of the equivalent sodium chloride (eNaCl) content or, if the composition is known, in terms of the relative salt contents and the actual salinity.

It is necessary that an accurate measure be taken of the water temperature, *at the time of the resistivity measurement.* This is not a problem, as a simultaneous temperature measurement or log can be routinely made with the resistivity measurement. In fact, a good temperature and differential temperature log, run simultaneously with the resistivity measurement, will provide an independent indication of layering and flows. Temperature and ion-type corrections were covered in *Introduction to Geophysical Formation Evaluation* in the chapter entitled "Water Characteristics and Ionic Conduction in Water". See also *Standard Methods of Geophysical Formation Evaluation,* in the chapter entitled "Spontaneous Potential, The Electrochemical or Diffusion Potential, Effect of Ion Types", for further discussion of ions in solution. *Standard Methods of Geophysical Formation Evaluation* also has further discussion of ionic solutions.

Because the total resistivity change from the bottom to the top of a body of water will be relatively small, the sensitivity of the resistivity or induction system can be increased to show small salinity changes. Small salinity and temperature changes also can be located by the use of a differential measure or readout circuit.

Presumably, a flow meter could be combined with these logs to determine any different flow directions of the individual layers or zones. As far as this author knows, this is not commonly done. It would, of course, require a good LandSat position reckoning of the surface vessel or moving platform and a vessel heave compensator (as would *any* of the wireline logs).

Standard oilfield or mineral logging equipment is often used for oceanographic measurements. Almost any of the resistivity or induction systems can be safely used. Frequently, oilfield equipment is not the most suitable nor the most accurate. It is, however, often the most available. Also, the field people of the major oilfield logging contractors are usually better trained and more knowledgeable than those of the usual small mineral logging contractor. Remember that oilfield instruments are designed to find hydrocarbons in sediments. They are not designed for any other purpose, even though they may be suitable for that purpose. In a large body of water, such as a lake or an ocean, the *type* of resistivity array used is relatively unimportant. A simple normal resistivity device can be as good as a focused electrode array or a modern induction log.

In a small body of water or a river, it would probably be wise to use a shallow depth of investigation device to avoid any errors from reading the bottom, top, or banks of the body. In any case, care should be taken to avoid "seeing" the hull of the boat or any wharf or pier. Note that three- or four-electrode lateral devices, the deep laterologs and the deep induction logs have a very large radii of investigation. If small flows, temperature differences, or small layers are to be detected, one of the focusing electrode devices, such as the Laterolog-3 should be considered. It might be even better to use one of the focused sidewall resistivity devices, because of its ability to read fine detail.

9.5.2 Laboratory Measurements

In addition to making resistivity logs, fluid samples can be taken and analysis made in the laboratory. Downhole fluid sampler takers can be considered. These are available in single or multiple sample systems. Also, there are fluid samplers designed specifically for oceanographic purposes. Some of these could be adapted to operate on a logging cable for remote control from the surface.

Oilfield mud testers can be used for oceanographic purposes. These mud testers are small, portable fluid resistivity measuring systems designed specifically for determining the electrical resistivity of small samples (~1 quart or liter) of drilling mud.

9.5.3 Fluid Testers

Laboratory and field testers ("mud testers" in oilfield parlance) are available in several forms. Most of them use electrode arrays and make resistivity measurements. The accuracy of these units varies widely from one model to another. Careful selection can result in surprising accuracy, which may easily be tested. The circuitry is usually standard logging-type four-electrode resistivity circuitry. The fluid containers are usually designed to be easily cleaned and are rugged enough for the oilfield. Many are capable of measuring small amounts of (~1 cm^3 or 0.06 $in.^3$) of semisolid mud (mudcake) and mud filtrate samples. They are convenient and available from oilfield supply houses and the major logging contractors. Their accuracy, however, may not always be great enough for scientific purposes and should be checked. Their drawback is that they require careful cleaning after each use. Conductivity meters are available from chemical and scientific supply houses, but often are not as accurate nor as convenient to use as the oilfield units. There are fluid resistivity testers (mud testers) that operate on the principle of the induction logging systems. The advantage here is lack of a metallic electrode in contact with the solution. Electrode-type units that have the array on a central, removable stem are probably better than if the electrodes are built into the wall of the cup. This is because the former is usually easier to clean. Scientific units are usually scaled in conductivity units. Almost all units use a fixed, calibrated volume of fluid, smaller than the radius of measurement, i.e., they are only calibrated to operate in a specific vessel.

Many of the simpler units use a two-electrode array. They apply a fixed voltage or potential across the electrodes and measure the current that passes through the fluid. The current is measured and usually calibrated in conductivity units.

Most of the more precise units use a four-electrode array. The two end electrodes are the current electrodes, A and B. The two interior electrodes are the measure electrodes, M and N. The current between A and B is held constant and the potential drop across M and N is measured. The unit can be scaled in resistivity or conductivity units.

Some of the more complex units use a two-coil induction array. These systems avoid some electrode problems. This type of system is used more frequently in oceanographic laboratories than in oilfield or mineral work.

Fluid testers may have a built-in temperature measurement device. Usually, this is just a mercury thermometer fixed in place in the solution. Some temperature measurements may be made electronically with a bridge circuit, a thermocouple, or a semiconductor sensor. There seems to be no inherent advantage or disadvantage to any of the devices, except convenience.

Regardless of the type of fluid resistivity tester used, there are several precautions that must be observed. Some of these precautions will depend upon the degree of accuracy desired:

1. Any meter used for scientific purposes (and more exacting than the demands of oilfield work) should be frequently checked with standard solutions. These solutions can easily be made using ion-free water and the salt mixture of your choice. Both water and salt must be accurately weighed. The solution must be complete. Accurate temperature readings must be made during the calibration.
2. In addition to frequent checking, it is imperative that the electrode array and the vessel or "cup" be clean. This is especially important if low salinity solutions are being measured. Some oilfield units have a cup which uses a squeegee on the end of the electrode array, so that the cup is wiped when the array is pulled out to empty the cup. This, however, is not good enough for careful laboratory work. In this case, both the electrode array and the cup must be carefully washed with ion-free water and dried and inspected before the next use.
3. Finally, *any* fluid resistivity measurement is meaningless without a simultaneous temperature measurement.

The units used in oilfield and mineral borehole work are *ohms meters squared per meter* (usually abbreviated *ohmmeters* or simply Ωm). Many surface geophysical and engineering units in the U.S. and Britain are scaled in *ohms per cubic foot,* which does not take the sample geometry into account. Scientific units are usually calibrated in *ohm-centimeters.*

9.6 Detection of Massive Discontinuities

Massive discontinuities are involved in many of the problems of nonhydrocarbon geophysics. As used in this text, these are structures that contain sharp lithological changes. The discontinuity may be a major fault, a metallic mineral deposit or body, a water-filled cave or mine, an air-filled cave or mine, a salt dome, or a basaltic or granitic pipe. In each case, the discontinuity forms an isolated, high physical contrast structure of some kind. There are and can be a number of techniques to isolate and define this type of structure.

9.6.1 Acoustic and Seismic Methods

Seismic methods can be used for detecting voids within the earth. Portable seismic units have been used effectively to locate near-surface limestone caverns. These caverns occur in coral islands in the Caribbean Ocean and contain potable water (personal communication). The portable seismic unit is small, about 16 × 12 × 9 in. (40 × 30 × 23 cm) and is battery operated. The noise source of these units is usually a sledgehammer and pad or a tamper. The limestone cavern represents a large discontinuity and is readily detected. Surface seismic methods have been used to locate extensively fractured zones at depth in the earth gas deposits, traps, faults, and most other lithological features.

Cross-hole and vertical seismic profiling (VSP) methods are used for detecting caves, lost mines, and mine drifts. The detectors and/or sources are used both on the surface and downhole. Downhole measurements are probably more effective on small targets, because of the distorting effect of the altered surface layers. These techniques have also been successfully used by the U.S. Geological Survey to differentiate between bituminous coal and scoria (burned coal).

Downhole acoustic logs are noted for their distinctive patterns when fractures are encountered. These "chevron" patterns are especially notable when the fracture plane is normal to the borehole. In massive rock, the focused resistivity devices can be combined with the downhole acoustic log to locate microfractured zones.

9.6.2 Tomography

Tomography is one of several techniques for constructing two- and three-dimensional images from the outputs of measuring instruments. In this particular case, the signals are from geophysical instruments. Other disciplines use these techniques under the same and different names. The most widely publicized use of the tomography techniques is the CAT scan of medicine. CAT stands for *computer-aided tomography.* The petroleum industry pioneered the use of tomographic methods under several different names; 3-D seismic and VSP are two of them. The first regular use of geophysical tomographic techniques appears to have been by Prospection Geophysique Francaise for the evaluation of dam sites.

Many tomographic studies use acoustic pulses or waves. This technique, however, can be applied to many of the measurement methods. Medicine uses acoustic sources and x-radiation. Geophysics uses acoustical sources and electrical resistivity/conductivity instruments. This method could be applied to any of the measuring methods that use a separate source and a receiver. Thus, scattered gamma ray density and neutron methods should be equally usable for smaller samples. Most of the geophysical methods, thus far, have been hole-to-hole (cross-hole). Surface-to-hole and hole-to-surface have also been used (VSP). Tomographic geophysics can involve any feasible means of acquiring a dense array of two-dimensional information along the surface and/or subsurface for multiple transmitter and receiver positions.

The tomographic technique requires a multiplicity of measurements to be made at many specific places of both the source and the receiver. VSP, for example, uses a sound source on the surface and an array of geophones in the borehole. The measurements are repeated with the geophone array at many locations throughout the borehole. Cross-hole measurements consist of repeated measurement with the source in one hole and the detectors in another at many different places throughout the holes. The more numerous the measurements and the more different locations of the sources and detectors, the better the results will be. Symmetry of the measured volume aids the quality, but it is not absolutely necessary.

Figure 9.8 shows a cross-section of a Canadian mineral deposit. The mean acoustic travel times for five of the included volumes are shown. The measurements were made in two boreholes and are indicated by the black circles or dots. The sources were on the surface at five locations. Figure 9.9 shows the probable ray paths and the mineral types. Figure 9.10 shows the tomographic contour section of Figure 9.9. Figure 9.11 shows a similar measurement with sources in many places on the surface and measurements made in two boreholes. Figure 9.12 shows the resulting velocity contours (Heinrichs, 1989).

The measured information is assembled to represent one or more planes of pixels, each of which has a shade representing a small range of signal values. If two or more planes are in reasonable proximity, a three-dimensional image may be formed. The holes and the planes may be in any spatial position. In fact, in addition to boreholes, mine drifts and tunnels are often used for the placement of the sources and the receivers.

Tomography provides a convenient, accurate, efficient, and cost effective way of gathering and examining geophysical data. It has the advantages of presenting the information in a form that can be in real time and easily read and understood. Software and hardware are available for acoustic use. Work is progressing on electrical parameter geophysical tomography.

Applications are foundation and site characterization studies, including the elastic moduli, fracture analysis, geological structure, and lithology. Their use is important in civil engineering projects such as dams, tunnels, waste disposal, canals, and underground storage. They are excellent applications for mineral exploration, ground water exploration and evaluation, hard mineral exploration, petroleum, and natural gas. They can provide dynamic monitoring of leachate fronts, fire fronts, steam fronts, and flood fronts. Horizontal arrays can be used for waste pond leakage, mine drifts, and water storage reservoirs. A combination of techniques can be used to locate fracture zones, mounds of water above the water table, and differing rock types.

The U.S. Bureau of Mines has computer programs for both straight ray and curved ray tomography. Curved ray analysis is required when the path of the ray is bent by velocity variations. Curved ray variations over 10% cannot be ignored. This bending is liable to occur near ore bodies and can be a factor of two or more.

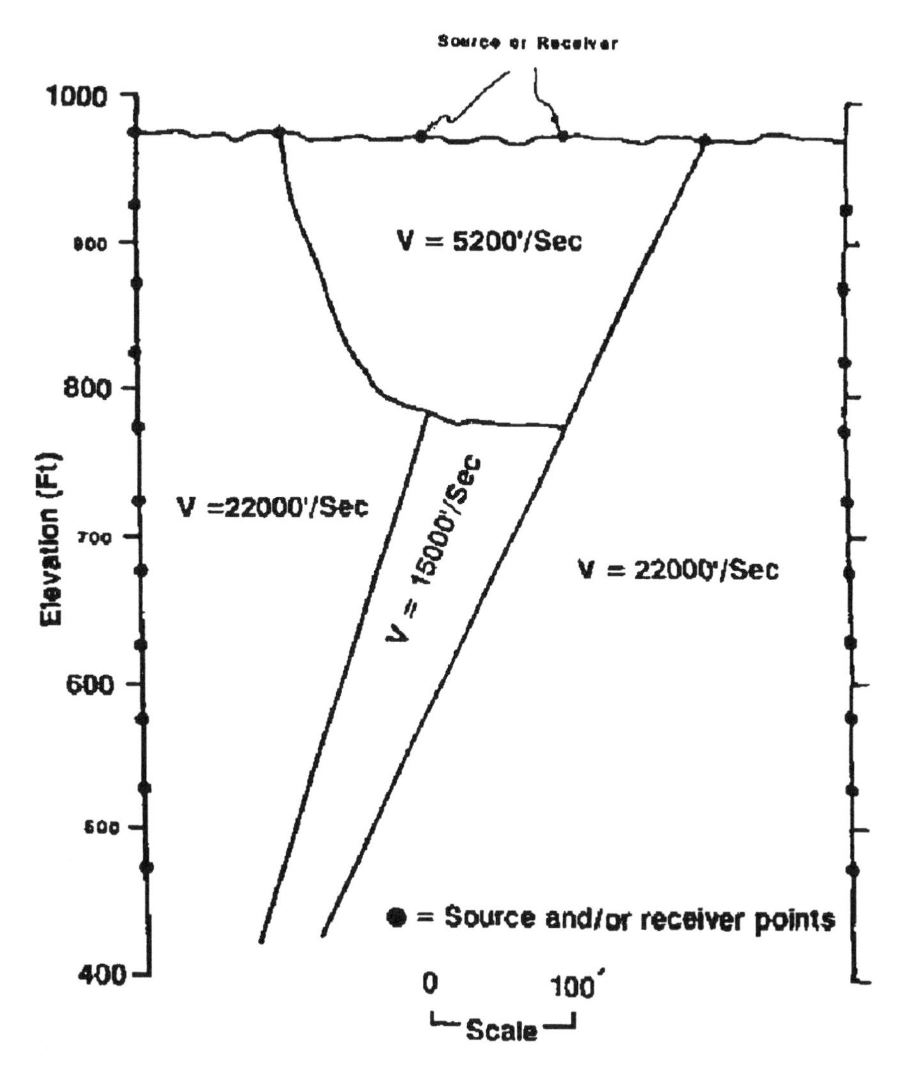

FIGURE 9.8
A tomographic setup and borehole readings from a surface acoustical souce (Courtesy of Canadian Geological Survey).

9.7 Mise-à-la-masse

One method for describing or defining highly conductive (compared to the surrounding material) bodies is the mise-à-la-masse (MALM) method. This method requires that the target be highly conductive and continuous. A good example is a massive sulfide body. The MALM method is usually applied

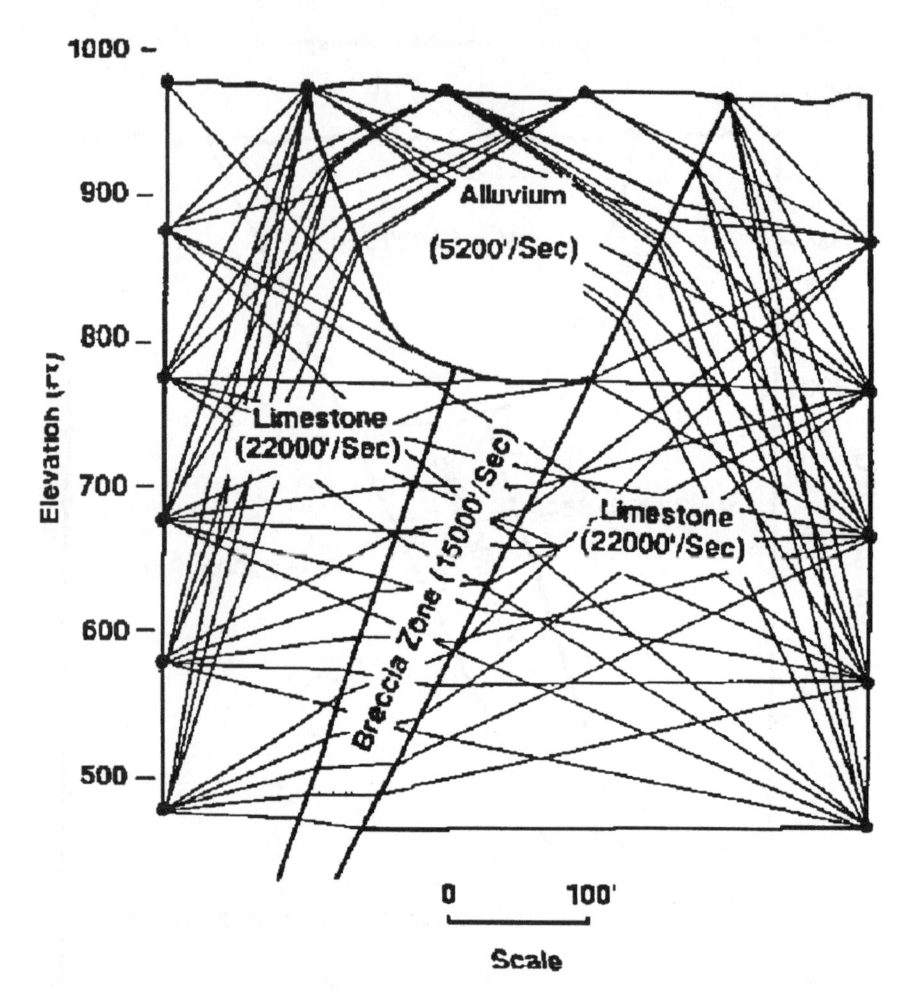

FIGURE 9.9
The ray paths generated in the previous setup (Courtesy of Canadian Geological Survey).

after the discovery of the body and is used to define the shape and extent of the mineral. Figure 9.13 shows the principle of the MALM method.

The MALM method uses the body as an electrode by flowing an electrical current into it. The fact that the body is highly conductive results in a more nearly equipotential condition over the surface of the body than in the surrounding formation. This, in turn, results in recognizable and predictable isopotential lines surrounding the body. Thus, the shape and extent of the body can be mapped by measuring the surface potentials. The surrounding potentials can also be read in nearby boreholes, mine drifts, mine shafts, and on the surface. The measurement can be of electrical potential or induced polarization (IP).

The MALM method has been used extensively in Canada on sulfide deposits. It was also used to locate water reservoirs under and near Paris, France.

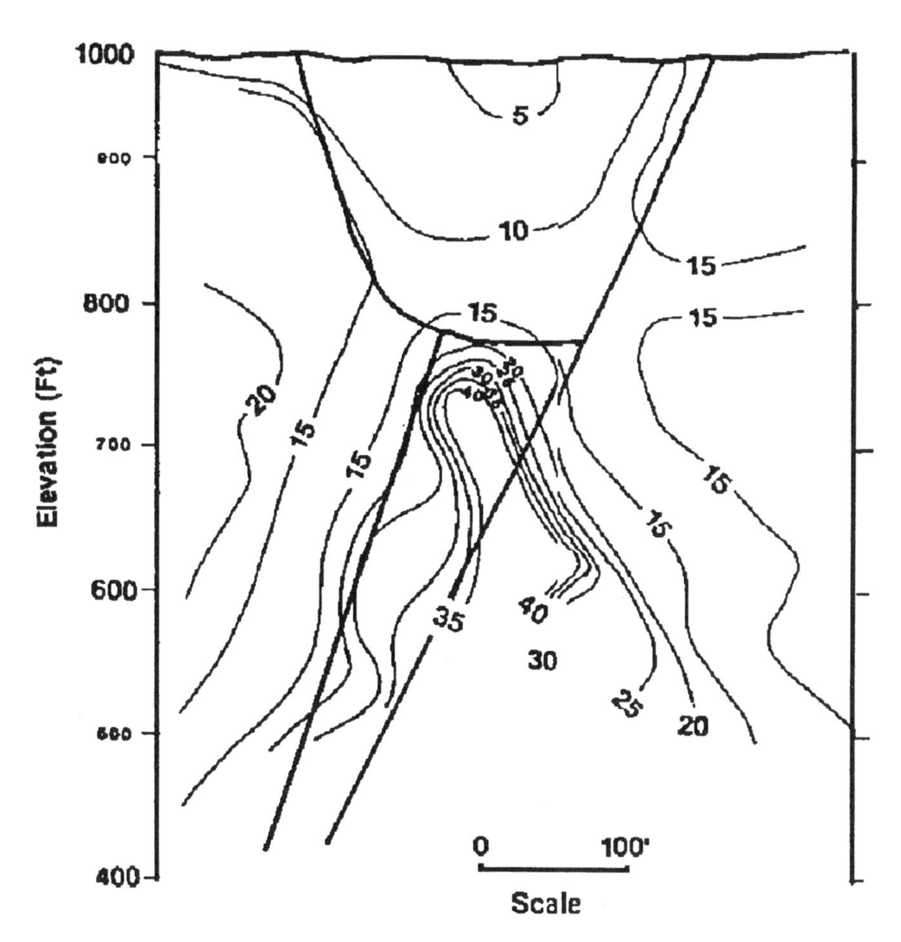

FIGURE 9.10
The tomographic contours determined for the previous setup (Courtesy of Canadian Geological Survey).

9.7.1 Reduction-Oxidation Method

Reduction-oxidation method (redox) is among the older means of locating massive metallic bodies. This method was used in the nineteenth century. It requires that a metallic body lie across the top surface of the water table. The oxidation of the portion above the water table, in the partially saturated zone, will cause electrical current flows between the top and lower portions. Detection consists of measuring the surface potentials. The clustering of the isopotential lines will pinpoint the mineral body.

This redox method is also used to locate "fish" in a borehole with standard spontaneous equipment. There was unpublished experimentation with the redox method to locate uranium "roll fronts" in the Wyoming Powder River Basin. The geochemical cell could be detected, but electrofiltration potentials

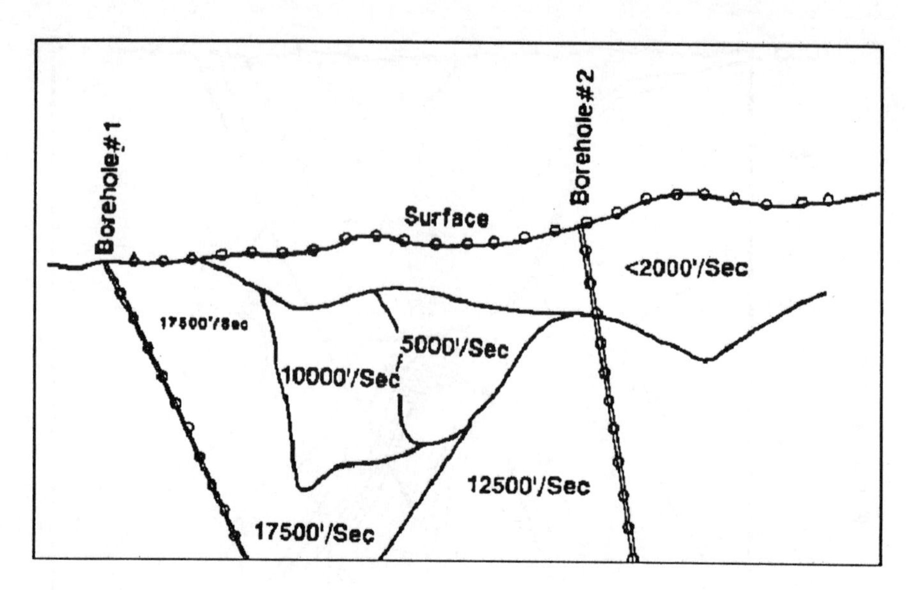

FIGURE 9.11
A tomographic setup (Courtesy of GEOEX).

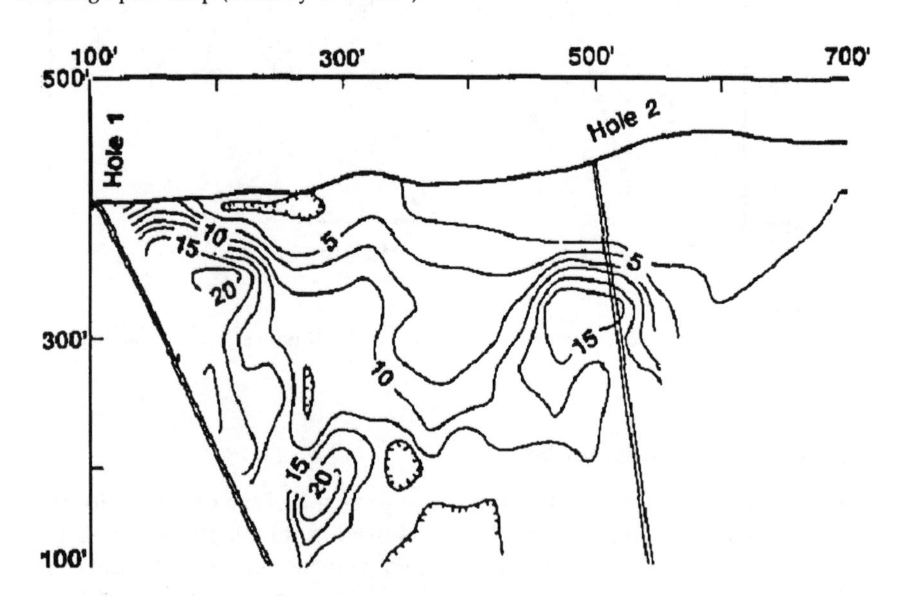

FIGURE 9.12
The contours for Figure 9.11 (Courtesy of GEOEX).

from the electrode contact fluids filtering down through the arid soil cause great interference. See also Section 9.9.2.1, Geochemical Cells.

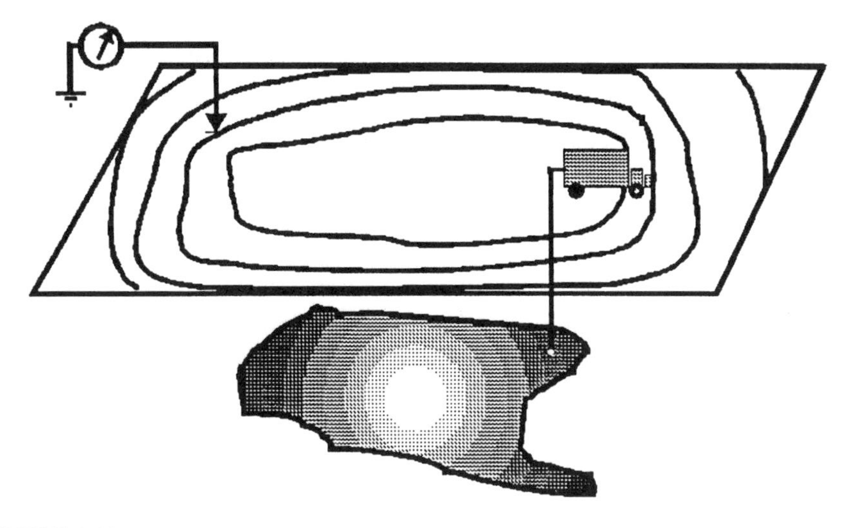

FIGURE 9.13
The principle of the MALM method.

9.7.2 Gravity

One of the older methods for locating massive metallic mineral bodies is the use of surface gravity measurements. Metallic minerals have densities of 1.5 to 3 times that of the surrounding zones, especially in sediments. Thus, it can be "seen" by the changes of the gravity values. This method, however, has a low resolution and the target body must be large. Large, low density bodies, such as coals, respond well to gravity measurements. Coals have low densities and range from 0.3 to 0.8 times the densities of the surrounding sediments. Water bodies also respond well.

Gravity surveys are routinely used in petroleum exploration to locate salt domes. Basaltic pipes and lava tubes should also respond to gravity measurements.

9.7.3 Magnetics

Magnetic field strength measurements from the air, surface, and downhole have been used for many years to locate massive sulfide and other metallic deposits. The density and electrical properties of these bodies suggest that a combination of gravity, seismic, and magnetic field strength on the surface could also be successful.

The induction log is probably the best known method used for downhole electromagnetic measurements. This device, even in its oilfield form has many uses. Some modern induction logging systems also determine magnetic susceptibility. This is done for correction purposes with some systems and for presentation as a separate log with others. It is an excellent device for salinity measurements in oceanographic work. It has the distinct advantage, for electrical measurements, in that there is no metallic contact with the target. Also, seawater resistivities fall within the best range for measurement by oilfield equipment. See Chapter 4 of *Introduction to Geophysical Formation Evaluation.*

9.7.4 Gamma Radiation

Massive metallic bodies and some fracture systems show some gamma radiation because of the precipitation of uranium compounds in their reducing environments. In the case of a sulfide body, airborne and surface measurements of gamma radiation show anomalous amounts of ^{222}Rn in the immediate vicinity of the body. The ^{222}Rn decay radiation is generally what is measured in airborne gamma ray surveys. This is apparently because the reducing environment of the sulfide tends to precipitate out water-soluble uranium compounds to a less soluble form. Fracture systems are often radioactive because water that contains radioactive materials flows through them. It is especially enlightening to run a gamma ray spectrographic log and a focused resistivity log in hard rock environments. In such a case, the presence of ^{238}U daughters suggests the passage of cold solutions. The presence of ^{234}Th suggests the passage of hot solutions. A full wave (P-wave, S-wave, and Stonely wave) acoustic recording should provide clues as to the clay, shale, or rock fracture fill material (see Chapter 2, "Photon and Particle Methods").

9.7.5 Surface Resistivity and Induced Polarization (IP)

Surface resistivity measurements are excellent for detecting lithological changes of an electrical nature. These methods are well proven and are readily available. If a surface resistivity is run, it is easy enough to also record the IP simultaneously that it is probably foolish not to do so. Further, these methods can be combined with downhole measurements to allow a tomographic presentation. These methods can easily locate discontinuities, but they are sensitive enough and have a high enough resolution to respond to lithological changes. They are particularly sensitive to metallic and sulfide deposits, and they can also pinpoint buried pipes, electrodes, or any other metallic material. They respond also to particular types of clays and any ion-exchange materials (see Chapter 1 in *Introduction to Geophysical Formation Evaluation*).

9.8 Engineering

9.8.1 Civil Engineering

Civil engineering has begun to use geophysical methods and geology in site evaluation, product testing, and in remedial studies. Geophysics, of course, is an ideal tool to use in civil engineering, because both disciplines are interested in the same parameters. The combination of civil engineering, geophysics, and geology is a natural one. The study and the use of physical properties of natural materials is the basis of all three disciplines. Even the historical side of geology is useful in engineering.

9.8.2 Structural Sites

The properties of the site chosen for a major structure or, even a minor one, can usually be determined with minimum cost and effort. The general site will be surveyed and the topography mapped. If the site is large, as for a dam or a major building, a gravity survey may be considered to locate any massive, subsurface anomalies. Seismic surveys can also be of value, especially if a tomographic presentation is calculated. These will show the possible faulting, voids, fracturing, and other such dangers to stability. If the surface layers are sedimentary, surface resistivity and IP are valuable for detecting and/or verifying the other surface methods. Gamma ray logs should also be run in sedimentary environments to assess the shale/clay content. Gravity surveys can determine the subsurface extent of massive rock formations.

The findings of the wide-area surface measurements should be verified with drill holes and logged with downhole tools. Density measurements can furnish valuable data about the surface soils, subsoil, and bedrock. Load-bearing values can be calculated from these readings. They will also detect local faults and fractures and will verify and support gravity and acoustic methods. The natural gamma ray log can give valuable information about the possible clay content of the upper sediments. It possibly may be able to call attention to movement of fluids within a fracture system. A Lithodensity log can identify the clay types. Resistivity measurements will determine the moisture content and salinity of the soil water, especially when combined with the spontaneous potential log. The neutron porosity log will give good porosity values, especially when used with the density log. If these are combined with a sample of the fluid, the ion content and the nature of the fluids can be assessed.

For very shallow investigations, the cone penetrometer is valuable in surface sediments. These measurements can be directly related to the load-bearing ability of the surface. Induction logging systems have been used to evaluate highway roadbed competency. In areas where the bedrock is too deep to use to support the structure, the best volume to float the structure can be determined with density logs. The amount of excavation can be calculated from the logged information. The practice, in an area such as Houston, TX, where the bedrock can be as much as 10,000 ft (3000 m) deep, is to excavate a mass of the soil equal to the projected mass of the building. A boat-like basement, or other substructure, is then built in the cavity and the building is built on that. This, in effect, floats the building on the sediments. The density of the upper layers of the soil are measured with oilfield borehole density tools. The water content of the soil is measured with the neutron porosity logging system. The combination of these two also allows an estimate of the clay content of the soil.

Electrical resistivity surface measurements combined with downhole resistivity and SP measurements are useful for economically determining the water content and salinity of the ground water. If a natural gamma ray log is also used, the shale and clay contents can be mapped to give an idea of the possible subsurface flow of the ground water. These same techniques can be used to locate possible salinity plumes down dip of a salt dome or a waste disposal site. They can also be used to assess the feasibility of ponds or lakes in any sedimentary area.

Acoustic and density logs are excellent for determining the condition of the rock upon which a dam structure is planned. This combination is particularly effective if a tomographic analysis is used. This combination is also useful for hard rock radioactive waste storage facilities. The tomographic analysis helps to pinpoint any anomalous features, such as faults and fracture systems, and estimate geological stability. It also helps to locate any hidden lithologic changes that may result in stress concentrations. Surface seismic profiling and VSP should also be considered for this application.

The combination of neutron and density readings are often used to determine concrete quality. The neutron measurement can determine the water content, including the water of crystallization. The density values can give a quantitative idea of the compressional strength of the concrete. This combination is used in highway inspection. Some oilfield contractors have offered this service in the past.

9.8.3 Right-of-Way Engineering

Highway and railroad right-of-way engineering are subspecies of site engineering. Most of the same problems are addressed. The big difference is in the distances involved. In both of these types of projects great distances aggravate the logistics of the work, and usually, time and cost restraints are common.

The combination of the density and neutron porosity logs is used to determine cement quality, after the concrete has set. This could also be used for roadbed and site evaluations. Hard rock areas, cuts, and bridge sites can be examined quickly and thoroughly with acoustic/seismic methods and tomographic presentations. The neutron/gamma ray spectrometer combination is a good way to determine the amount of soil water and clay content. Some investigation has been done in Massachusetts using the induction log to determine roadbed packing. The cone penetrometer is also an ideal tool to evaluate roadbed quality before construction and paving.

Low level visual aerial photography is already being used for initial planning. Satellite imagery can be used to anticipate topography, forest, and routing problems. Infrared imagery can help locate underground water and zones that have been disturbed in the past. Visual imagery can give a good idea of the topographic details and possible obstacles that have been omitted from the maps.

9.8.4 Mining

The mineral mining industry has been using geophysical methods for many decades. Such use could be expanded greatly. Gamma ray, detailed resistivity, and SP measurements are now routinely run on many projects, especially on uranium and evaporite projects. Hundreds of low cost, shallow boreholes are drilled and logged by fleets of truck-mounted drilling rigs and small logging trucks. Large areas are examined by gravity measurements, airborne gamma ray and magnetics, and often with seismic profiling and VSP. Rippability of the overburden, for open pit mines, is predicted by density and gamma ray logs. Slope stability is estimated with gamma ray, density, and neutron logs. Underground mines even occasionally use the modulated currents in the ground for telephone communication between underground and the surface. Density logging has been routinely used to monitor the progress of cementing vent and access shaft linings.

9.8.5 Real Estate Evaluation

Real estate evaluation can be handled in much the same manner as the site evaluation in the previous section. The main difference is that the interest in real estate evaluation is mainly in the very shallow and surface layers. Here, mineral exploration practice is ideal since it, too, is concerned with shallow investigation. Of prime importance are drainage patterns and load-bearing characteristics. The electrical resistivity, SP, natural gamma ray combination can be used for this. These readings can easily be combined to display a two-dimensional tomographic picture of the site. Shallow seismic measurements may also be of use. If immiscible fluids are involved, as on oil storage properties, ground penetrating radar (GPR) may be useful for determining oil flow patterns and near-surface contamination. Geophysical measurements should be backed up by chemical analysis of soil samples. The use of low level and satellite imagery is a valuable, inexpensive tool.

9.8.6 Archeology

The use of surface geophysical methods is an excellent way to nondestructively explore an archeological site. Shallow seismic measurements, electrical resistivity, GPR, IP, and surface SP mapping can give a good picture of the possible value of a site well before any excavation is done. It is also possible that a surface natural gamma ray map could be effective, because of the concentration of uranium compounds in the reducing environment of organic materials. This field is already using satellite infrared imagery.

9.8.7 Waste Disposal

Waste disposal has already discovered the value of geophysical techniques. Potential sites for high level radioactive waste disposal have been thoroughly logged, in spite of objections from tradition-bound engineers. Combinations of surface and borehole techniques are routinely combined in tomographic presentations to examine potential sites and examine these sites after disposal has taken place. Monitor wells in land-fill installations are routinely checked with downhole television to locate possible leaks, damaged or collapsed casing, and undesirable growths. Injection techniques and the logging support are used for deep formation liquid disposal. Fluid injection and dye and radioactive tracers are used to determine flow patterns.

9.9 Metallic Mineral Exploration

There are many mechanisms that transport, deposit, remove, and concentrate mineral deposits. Many of these mechanisms can be detected with geophysical methods by their actions, their results, and by the traces left behind. Further, the mechanism probably required a particular set of circumstances in order to operate. These requirements may be a temperature change, a particular pH, or a water flow. These things can be detected by geophysical/geological means, if one is alert and reads the signs properly. Every indication on a geophysical log or map has a meaning. It is up to the user to decipher these indications.

9.9.1 Transport and Deposition Mechanisms

9.9.1.1 Oxidation-Reduction Phenomena

Oxidation-reduction phenomena, also known as redox phenomena, are common mechanisms for the deposit and transport of sedimentary minerals. One, oxidation or reduction, cannot occur without the other. If something is oxidized, something else must be reduced. Redox occurrences are very common. Virtually all of our modern energy production occurs by means of oxidation. The prime energy source is our sun. Virtually all of our energy comes from it. Nevertheless, the energy from the sun is stored and expended by redox mechanisms. All life processes are redox-based. A large portion of our electrical and mechanical energy derives from redox reactions. Decay processes and most alterations are redox actions and some mineral transports and depositions depend upon redox processes.

It is the last of the above list of redox processes we will address first. These form an important process in sedimentary mineral deposits, particularly.

The process of reduction is that of gaining an electron. The electron must come from somewhere, and that somewhere is from oxidizing something else. The process of oxidation is that of an atom losing an electron. For example, carbon can be oxidized by reducing an oxygen atom, with the release of some energy, E (heat), to become a molecule of carbon monoxide:

$$2C + O_2 \rightleftharpoons 2CO + E \tag{9.1}$$

Similarly, iron oxide is reduced in a furnace by adding heat and oxidizing carbon:

$$2Fe_2O_3 + 3C + E \rightleftharpoons 4Fe + 3CO_2 \tag{9.2}$$

This type of reaction is important in many mineral transport and depositional processes because when a redox reaction takes place, chemical and physical properties change. This was made obvious during the exploration for uranium in sediments in the U.S. from 1940 through 1980.

Sedimentary uranium occurs in deposits that are (or were) geochemical cells popularly called roll fronts. Many minerals, including some of those of uranium, copper, and silver, have the property of being more soluble in water when in an oxidized form than in a reduced form. For example, cuprous chloride (a reduced form) has a solubility in cold water of 6.2 ppm and cupric chloride, a more oxidized form, has a solubility in cold water of 6.8×10^5 ppm.

Gas seepage of methane, carbon monoxide (both good reductants), carbon dioxide, and nitrogen in New Mexico frequently have halos of metallic minerals around them. Concentrations of silver were high enough around some of them to result in a serious evaluation concerning mining them.

9.9.2 Relation to Geology/Geophysics

Other minerals besides copper and uranium display this action. The U.S. Department of Energy had a photographic slide that they used for several years to lecture about uranium roll fronts. The near-perfect roll front, however, was a copper deposit, not a uranium one. The process is a common one.

The process which appears to take place in roll fronts has several prerequisites:

1. There must be a proper source for the mineral, in the first place. It may not still exist, but it must have been there during the initial stages. The source rock in many Rocky Mountain deposits was weathering granite outcrops. As the granite decomposes, the uranium component mineral leaches out, forming a very dilute oxidized solution of a uranium compound.
2. There initially must be a permeable sand outcropping downflow from the source for the dilute solution to flow into.
3. The sand layer must be confined above and below by impermeable shale layers to form a conduit down which the solution will flow.
4. While the near-surface portion of the sand may be oxidized, the lower, down-dip portion must be neutral or reduced. This is a normal situation for formations well below the surface.
5. At some location, down dip, there must be a reducing material. It may be dispersed or it may be concentrated at one location.

The dilute solution of the mineral will flow into the sand outcrop and continue slowly down dip. When the solution hits the reduced zone, the solubility of the dissolved mineral will decrease and the mineral or some of it will precipitate out. Therefore, the more oxidized form of the mineral will be reduced and become more reduced (less oxidized) at the redox interface.

Subsequent oxidized solution will flow into the precipitated, reduced mineral body, oxidize the back of it, carry it forward a little into the reduced environment where it will be reduced, and deposit it farther forward. This will be a continuing or repetitive process that will cause the geochemical cell to slowly creep forward and increase in size and concentration.

As it moves along, the concentrated, mineralized main body is in contact with the shale-sand interfaces above and below. The mineral will slowly diffuse a short way into the shale and be left behind because the oxidizing solution cannot reach it. Also, because the mineralized solution will be heavier than the virgin fluid in front of the cell, the enriched solution will have some tendency to flow under the new waters.

Thus, we see the classic roll front shape of the geochemical cell. Other minerals, besides uranium and copper can form the same kind of cell. We have a good picture of the uranium cell because thousands of gamma ray logs mapped it. The copper cell had a bluish color and was in a road cut, thus we could see it.

9.9.2.1 Geochemical Cells

Many metallic compounds change solubility, depending upon there redox (reduction-oxidation) state (Garrels and Christ, 1965). This can result in transport over great distances in the oxidized state and subsequent deposition upon becoming reduced. This is the mechanism of the uranium roll front. It also seems to apply to copper, silver, and several other metallic minerals investigated by the author. It appears common in near-surface sediments and some evidence indicates the possibility of the same mechanism operating in many hard rock mineral deposits.

In addition to the redox effect, many minerals, in deposit, are often accompanied by various degrees of radioactivity. The radioactivity of uranium compounds is well known, as is the radioactivity of shales. This radioactivity suggested the roll front idea and allows easy location of many mineral deposits.

A uranium geochemical cell (a roll front) has a characteristic pattern. This pattern is duplicated in several other mineral-type redox deposits. The Atomic Energy Commission (now the Nuclear Regulatory Commission) used a photograph of a roll front, in lectures, for many years. A road cut had followed the path of a roll front, so that an excellent picture of the front was presented. The mineralization, however, was not uranium, but copper. On a silver project in Oklahoma, this author had a detailed analysis made of a redox interface. A number of metals were involved. The sequence of metal ion types precisely followed the electromotive series, as published in handbooks.

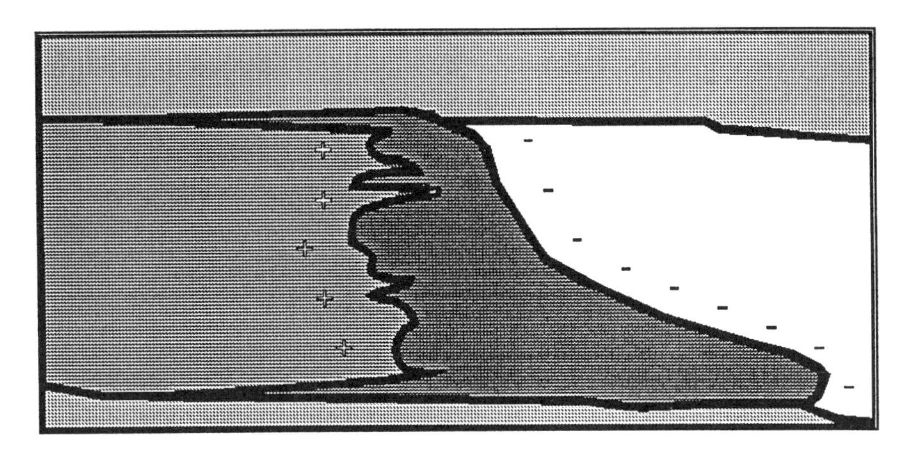

FIGURE 9.14
A diagram of a roll front geochemical cell.

A typical uranium roll front will occur in a permeable sand conduit formed by an impermeable shale below and usually an impermeable shale above (see Figure 9.14). The usual zones, determined by the geologists, are:

1. The uranium roll front is typical of the conduit-type of geochemical cell. It results from the flow of oxidizing waters, usually from the surface or it may enter at an outcrop. The permeable portion of the

outcrop acts as a conduit if it is bounded on the top and bottom by impermeable shales. The water will flow down dip, through the permeable bed (sand). Because these waters may contain dissolved oxygen and other oxidizing agents, they will alter and dissolve any of the susceptible minerals in their path. Over time, this zone becomes highly oxidized and depleted of the soluble minerals. It frequently has a characteristic light color — tan, light brown, yellow, or orange. The top and bottom shales may be similarly affected because of the migration of the oxidizing agents into the shale. This is called the *altered, remote barren zone.*

2. As the fluid progresses, it becomes richer in the soluble minerals. Some of these minerals will migrate into the bounding shales, in spite of their probable very low permeability. Thus, they will show anomalous trace mineral amounts near the boundary. Likewise, any shale streaks or lenses in the sand will have some small amounts of the mineral diffuse into them. The sand, itself, however, will have little of the dissolved mineral in it. Its color will still be light. This is the *altered, barren zone.*
3. Further down dip, some of the dissolved mineral begins to remain in the sand, near the boundaries. The bounding shales will also have some diffused mineralization. The center and near center of the sand will still be almost barren of dissolved mineral. The color of the sand will still be light yellow to red. The shales may be gray, except at the boundaries, at this point. This is the *altered, near barren zone*. Here, the solution still contains all or most of the mineral it has acquired up dip.
4. After the near barren zone, the fluids may begin to encounter some of the reduced sand, especially if there is an excess of a reducing agent present. The reducing agent may be organic trash and humic acid, sulfides, or some other reducing environment, perhaps even bacteria. This reductant will react with the oxidized, dissolved mineral, change its solubility, and cause it to begin to precipitate out. The vertical pattern of the deposition will be quite irregular because of the variations in the permeability of the sand. These will be emphasized by the amounts of deposited mineral. The color, here, will vary from light to dark and be streaked, as the redox state and the amounts of precipitated mineral vary. This is the *interface zone* and is the most active, chemically. In addition to depositing out the mineral, the solution will also pick up some of the previously deposited mineral, as the reductant is depleted by new action. This re-solution will be moved forward into the main body interface.
5. Finally, the mineral body will exist in down dip contact with the interface zone. The bulk of the mineral will be found here. The color will be dark gray, dark brown, or black. The waters past this

> point will have much less mineralization. However, at the bottom front of the cell, because of the slightly higher density of the mineralized solution, a long nose will form on the lower boundary, in the *reduced zone*.

This pattern is followed by other metallic minerals that can be found in geochemical cells. A diagram of a typical geochemical cell for uranium is shown as Figure 9.14. The pattern of a roll front can be seen clearly on a suite of downhole logs. It appears (verbal communication) that its location can also be detected with the proper surface methods. The original detection of this type of geochemical cell was with the gamma ray downhole log and was for sedimentary uranium. These methods can be applied for the search for other metallic mineral types, with some modification.

The SP measurement is affected by the redox state of a zone (please refer to *Standard Methods of Geophysical Formation Evaluation*). Information about the relative redox state in a borehole can be obtained from examining the slope of the SP baseline approaching or leaving a geochemical cell or any other redox change, such as surface oxidation. On the surface, the SP can be used to spot the cell, if certain precautions and restrictions are observed. This method was first used in the late 19th century. The electrodes must be a stable type, such as silver-silver chloride construction. The initial field tests used copper-copper chloride electrodes. Allowance must be made for and electrofiltration potentials arising from the possible flow of fluids from the electrode site into the soil, especially in arid surface soils. Close spacing on a predetermined pattern will allow mapping of the isopotential lines which are the result of current flowing from one part of the deposit to another, through the resistance of the ground.

The types of geophysical methods to be used on these deposits will depend upon the mineral being sought. The initial work was done on sedimentary uranium deposits, especially in the Powder River Basin in Wyoming. However, minerals, such as silver and copper will follow the same patterns. Probably there are several others, as well. Several ore grade silver (and other minerals) halos were found around natural gas seepage in New Mexico.

The early uranium effort used surface geology, exclusively, to locate likely areas for examination. Permeable outcrops were sought, especially if they were near granitic deposits. The sands and bounding shales were checked with a hand-held radiation detector. Airborne and car gamma ray surveys were run to try to detect the presence of radon (^{222}Rn), because ^{222}Rn is a daughter product of uranium (^{238}U). Surface helium detection was sometimes used, because ^{222}Rn decays by the emission of an alpha particle, which is a helium (^{4}He) nucleus. Another method was to place a pattern of "track etch" cups on the field. The alpha particle tracks of individual cups were counted and plotted, by field location, on a map. Isonumeric line highs tend to outline the deposit.

Boreholes were then drilled on a grid to locate the outcrop sand. The probable depth was estimated on the basis of the known dip in the region. The color and type of the cuttings were recorded to determine the probable position, with respect to the redox situation. The initial spacing was fairly coarse, until the redox interface was bracketed. Initially, a gross count gamma ray log was the only log run in the boreholes. Later, combination tools were available to record the gamma ray, SP, and single point resistance curves. The latter became the standard suite of logs for uranium. Later, other logs were added to the suite as increased knowledge and the situation demanded.

The gamma ray log showed the presence or absence of radioactive mineral. Radiation levels, above the normally expected formation level (background radiation), were looked for to locate the cell and its passage. The SP curve was initially used to distinguish the sand and shale sequences. The resistance curve was used to help identify the sand/shale sequences and locate thin bedding within the sands. The combination of the signatures was used to identify each zone, with respect to the known geological column of the region. The shape of the signatures was used to locate the borehole with respect to the geochemical cell. The gamma ray log was calibrated to allow the calculation of the probable mineral grades in the anomalous zones. Dips were calculated from logged depths and topographic maps.

Core samples were taken in strategic boreholes to verify, in the laboratory, the equilibrium state of the mineral. The cores were first run through a core gamma recorder to allow correlation with the downhole logs. Samples were then subject to chemical and X-ray analysis to determine the actual uranium (^{238}U) content. This was then used to correct the mineral grade values calculated from the gamma ray log. The initial results of the drilling and logging allowed drilling on closer spacings and homing-in on the geochemical cell. The above procedures were repeated in this closer spaced grid.

There are many features that can be added to this exploration picture. A neutron porosity log was added, later, to allow the calculation of porosity and estimation of shale content in the presence of the high radiation. Density logs were often run to help identify the alteration of the zones. Of course, the present gamma-gamma density systems are useless in the high radiation zones. The density was also used to locate the water table.

In about 1960, the PGT system was introduced to the field. Unfortunately, the system was complicated and relatively expensive. It was highly successful, however. This system used a high resolution, downhole, gamma ray spectrograph to detect the presence of protactinium (^{234}Pa). ^{234}Pa is the third daughter of ^{238}U and is separated from it by about 24 hr. Thus, the presence of ^{234}Pa indicated the high probability of the presence of ^{238}U.

9.9.3 Other Mechanisms

There are many mechanisms for the transport, solution, and deposit of minerals. An alternate to the traditional exploration method of looking for the mineral would be to look for an environment that is or was favorable for supporting one of these mechanisms. This can be done with geophysical means.

Then, when a favorable mechanism is identified, the probability of finding a resulting mineral deposit is greatly enhanced.

9.9.4 Acid-Base Reactions

Acid-base reactions have effects upon some compounds and the result is similar to the redox reaction. Some compounds are soluble in a low pH environment and less soluble or insoluble in a high pH. A central Australian sand (low to neutral pH) has a high concentration of uranium on and around buried boulders of limestone (high pH).

9.9.5 Thermal Changes

An obvious mechanism that can be observed in action around undersea volcanic vents is the precipitation of minerals when the hot volcanic solutions hit the colder ocean water. Hydration-dehydration is another mechanism. We can see this in action with the stalactites and stalagmites in caves. We can see the end result in the extensive beds of evaporite in many places in the world.

9.10 Coal

9.10.1 Surface Methods

Various geophysical methods, other than borehole logging, can effectively be used in exploration and evaluation of coals. All coals have relatively low densities, compared to the sediments in which they are usually found. A few geological materials have densities that fall into the range of coals, but not many. Because economic deposits of coal must be fairly thick and wide in areal extent, several surface methods suggest themselves as technically and economically viable in the search for coal. As far as this author is aware, most of these are seldom used.

The most obvious method of surface exploration for all forms of coal appears to be gravity surveys. The drawbacks appear to be the large volume encompassed by the gravity measurements and the resulting lack of sharp boundary definitions. The contrasts with the usual formation materials are great, even with respect to anthracite. These contrasts appear to be on the order of 2.4 for lignites to 1.3 for the most dense anthracite. Their volumes can be measured in terms of cubic miles or kilometers. Bed boundaries will show up, of course, as tight spacings of the isogravity lines.

Seismic methods, also, should be effective in the search for coal. While the contrasts are not as great as those of oil or gas deposits, their extent will be about the same. Acoustic travel times of anthracite are about 90 to 120 μs/ft (295 to 395 μs/m), bituminous coals are 100 to 140 μs/ft (328 to 459 μs/m), and in lignites, 140 to 180 μs/ft (459 to 590 μs/m). In comparison, the travel times for waters, in those depths, are 207 to 192 μs/ft (679 to 630 μs/m) and for shales, are 60 μs/ft (197 μs/m).

Coals are organic residue. They have been compressed, heated, and otherwise altered. The probability is high that the only substantial radioactivity in a coal, with a few exceptions, was from carbon-14 (^{14}C). Carbon-14 has a half-life of 5730 years. Acquisition of ^{14}C ceased when the plant stopped growing, which was about 1700 half-lives (or more) ago. Thus, the amount of residual radioactivity is minuscule. Essentially, the coal is usually almost nonradioactive. However, the density log should always be accompanied by a gamma ray log. Thus, coals have low radioactivity and low densities. Further, the lower grades of coals have medium to very high electrical resistivities. These features suggest detection by the gamma ray, density, resistivity, gravity, and acoustic systems.

The density system is an excellent detector for the presence of coal beds. It should, however, always be used with other methods. A density tool, as used in the petroleum industry and most of the mineral industry, requires intimate contact with the wall of the hole. This is usually available in a good coal bed, regardless of the coal grade. Because coal is mostly nonpermeable, there is no mudcake; therefore, a compensated density tool is not needed. Because this is typical of most mineral logging, the compensated density system is normally not used on nonpetroleum projects. A gamma ray device is usually part of a density system.

There have been promotions by some logging contractors to detect coal beds by using a simple gamma ray tool with a small gamma ray source on the bottom. This actually works. It is only qualitative, however, and the actual positions of the bed boundaries are always in question (by as much as the spacing between the source and the detector). It is a method that can be fielded in a hurry. It should be followed up by a suite of quantitative logs.

Coals are usually laid down in layers and thus often are laminar. If the coal sloughs, it can occasionally result in the surface of the borehole, in the coal bed, being rough or even caved. This condition will show up on the caliper, which should *always* be run with a density log. Again, a caliper device is normally part of the density system.

There are exceptions, of course, to the normal nonradioactivity of coals. These especially occur when the coal is invaded by solutions of formation waters. These waters may carry clays, uranium, potassium, and/or thorium compounds and their daughter products. The invasion may take place during the deposition of the original organic matter. The layers of coal may be or have initially been permeable, allowing the invasion of small amounts of radiation-bearing formation water; or, subsequent fracturing may have allowed formation waters to flow through the coal bed. Usually, though, the coal beds are characterized by their very low radioactivity. They normally stand out on the gross count gamma ray logs because of their low gamma radiation. On the other hand, the surface and airborne gamma ray systems are normally useless for coal exploration. This is because the overburdens are thick enough that the amount of gamma radiation originating from the coal, even if it is "hot" (radioactive), is negligible. Thus, other surface exploration methods must be used. Surface seismic and gravity surveys suggest themselves because of the low density ranges of *all* coals (<1.0 to 1.8 g/cc).

The amount of clays or shales in coals are most of the ash content. Some carbonate and/or sand may appear, but the ash seems to be mostly from the clay minerals. This means that the gamma ray curve becomes an excellent way to estimate, *in situ*, the ash content of a coal bed. We find that the hard coals — anthracite and metallurgical coals — have very low clay (ash) contents. Bituminous coals generally have clay contents about that of very low radioactive sandstones or lower (typically 0.1 to 1 API). Lignite and peat have relatively high radioactivity and ash content, compared to the other coals. These may be 20 to 40 API or more. Some contractors publish charts which attempt to relate ash content to the relative gamma ray log readings. These should be considered qualitative, at best.

In addition to the caliper curve being necessary for the density log, it can be very useful when used in conjunction with the gamma ray log. Used thus, it can give an indication of the quality and condition of the shale roof and the location of the floor shale. Used to evaluate the roof, it can also accurately indicate the vertical extent of the coal, as it grades into the shale. The gamma ray curve can also give clues as to the fill of the partings.

A high scale, high resolution resistivity curve can accurately locate partings in a coal. The higher the resolution of the resistivity, the finer the parting to be detected. Because most of the focusing electrode resistivity systems are essentially single point curves, the parting boundaries will *always* be indicated by the inflection point of the resistivity curve. An expanded depth scale, on the log, is a great aid in picking these bed boundaries. In addition, the mineral-type focusing electrode resistivity systems are generally designed to read higher resistivities than do petroleum systems. There is small need for a resistivity reading, in petroleum efforts, above about 2000 Ωm. On the other hand, a bituminous coal can have electrical resistivities up to 20,000 or 30,000 Ωm. There is also an indication that the resistivity of a bituminous coal is related to the quality of the coal. One coal logging contractor, in Alberta, Canada, routinely runs a system which he says is usable to 30 kΩm. The resistivity system is, of course, valuable in the noncoal parts of the borehole. Here, it helps identify the stratigraphic sequences and accurately locate the target, with respect to the local geological column.

Surface resistivity surveys can be quite valuable in exploration for bituminous and, perhaps, anthracite beds. The resistivity contrast of anthracite coals is not as good as that of bituminous coals. They will typically be 100 to 10,000 Ωm for bituminous coals and 0.5 to 20 Ωm for anthracite. Density and gravity methods will respond to the large areal extent: the 3:2 density contrast of anthracite, and the 3:1 density contrast of the bituminous coals. The large horizontal extent of these beds, their characteristic resistivities, and their relatively shallow depths make these surface systems ideal for exploring for and mapping out coal beds.

The moisture of a coal will exist in the coal itself and also in the clay of the partings within the bed. The neutron porosity system can be used to determine the total amount of moisture present. It should be remembered, however, that the neutron system does not distinguish between the water and the liquid hydrocarbon within the coal. The liquid hydrocarbon forms part or all of the volatile content of the coal.

When the four curves — the resistivity, density, gamma ray, and the neutron porosity curves — are run through a coal bed, they can be combined in a cross plot to approximately determine the four important parameters of a coal. These four parameters are the heat, ash, moisture, and volatile contents of the coal. Because a two-dimensional cross plot is generally the most practical and easiest to use, the four curve values can be represented by two ratios. Because both the resistivity and the neutron respond to the moisture content, the gamma ray responds to the ash content and the density to the carbon content, the two ratios might be

$$y = \frac{R}{GR} \tag{9.3}$$

and

$$x = \frac{\rho}{N} \tag{9.4}$$

Thus, one might envision a cross-plot diagram such as shown in the Figure 9.15. Various means might be used to enhance such a presentation. Weighting factors can be applied, other combinations in the ratios might be used, and a three-dimensional presentation can be used with the help of a computer.

The drilling muds used in a coal/mineral exploration project are minimal. That is, they are designed for economy rather than for borehole competence. While this usually is often a false economy, it does result in the floor and roof of the coal bed showing characteristic caving. This phenomenon is partially due to the salinity differences between the drilling mud and the formation. A drilling mud salinity lower than the formation salinity will cause the shales to slough because of osmosis. A typical response of the driller is to add more bentonite to the drilling mud, which does not solve his problem and makes logging difficult and hazardous.

The logged curves will usually give excellent indications of the coal bed boundaries. At the floor of the coal, the shale boundary is usually sharp. The roof boundary, on the other hand, is frequently a gradual transition from coal to shale. Thus, the problem in estimating the coal amount is one of determining where the roof shale content becomes so high that the coal is uneconomic to mine.

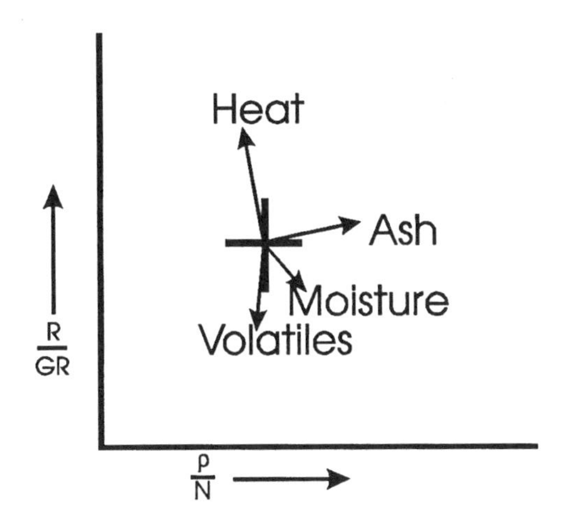

FIGURE 9.15
A hypothetical coal cross plot.

The floor shale is less affected than the roof shale. This is because the floor shale was in place before the coal-forming material was deposited. Typically, its signature is as shown in the idealized drawing seen in Figure 9.16.

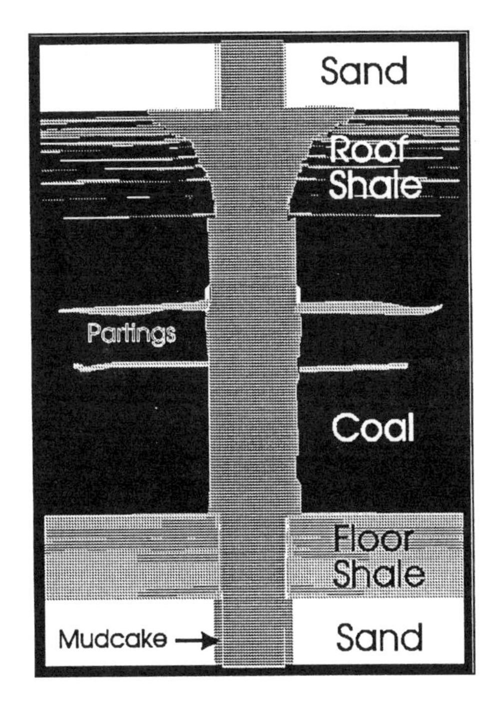

FIGURE 9.16
A coal zone, showing the foot and head shales.

The gamma ray curve will show the transition from coal to shale, quantitatively. The caliper curve will show the sloughing and partings. The focusing electrode resistivity will show the partings, roof layering, and boundaries, as well as some of the quality of the coal. The neutron curve will show the moisture and volatile content of the coal and compare them to the floor and roof shales.

The resistivity of a coal bears a quantitative relationship to the quality or grade of the coal through its ash and carbon content. The peat, lignites, and brown coals have high water contents and high ash contents. Thus, these have low electrical resistivities and densities. Because the salinity of the waters in most coals is low, however, the resistivity is lowered by their shale content as well as their moisture content. Thus, a lignite may have a resistivity in the range of 1.0 to 2000 Ωm. This is often higher than the neighboring shales and sands, but still relatively low. Their densities will be around 1.0 to 1.2 g/cc. Neutron logs will show high water/volatile contents (30 to 50%). Gamma radiation will be moderate because of the shale content. Sulfur contents are usually moderate to high.

Bituminous, soft, or black coals have high resistivities and moderate densities. This is because of their form of carbon, their relatively low ash and moisture contents, and their volatile content. Of course, any partings that probably will be filled with shale and water will lower the average resistivity. Any nonparted zones will have resistivities much higher than the lignites and perhaps as high as 20 kΩm or higher. In general, the higher the resistivity, the higher the grade of the bituminous coal. Densities of bituminous coals will be from about 1.1 to 1.3 g/cc. Neutron thermal neutron capture cross sections are about 1.54×10^{21} b/cc. Gamma radiation will usually be extremely low. Sulfur contents may be low to high.

Anthracite, shiny, hard, and metallurgical coals again will have low resistivities but higher densities. The low resistivities are the result of the form of carbon in the coals. Substantial electrical conductance occurs in the carbon itself. The higher densities are also due to the form of carbon and the low ash or shale content. Electrical resistivities of anthracite range from about 5 Ωm to as low as 10^{-3} Ωm. Densities will range from about 1.5 to 1.8 g/cc. Neutron capture cross-sections will be about 1.1×10^{-21} b/cc. The gamma radiation is usually negligible. Sulfur contents will generally be low to very low.

9.10.2 Other Geophysical Techniques for Coal Exploration

There are many possible uses of the geophysical techniques that have not been covered here. Environmental studies and hydrology both use these methods now. Oceanographic studies have been using oilfield contractors and their petroleum-oriented systems to a surprising degree of effectiveness. They would certainly benefit from the use of systems more closely suited to their needs. Forestry, animal migration studies, and urban planning are already benefiting from the use of infrared satellite imagery.

Sampling devices should not be forgotten. Unfortunately, downhole sampling systems frequently require large boreholes. There are, however, sidewall soft formation samplers that are operated with the drill pipe. Commercial water wells could certainly use the available sidewall sample guns and fluid samplers that can take multiple samples. Hard rock sidewall core samplers are large. When the hole is available, they are reliable and inexpensive.

9.11 Analysis

9.11.1 Cross Plotting

Cross plotting methods of analysis of logged geophysical data are popular methods that have been used for many years. They were suggested by John Wahl of Schlumberger Research in about 1970. The method has never been very popular in nonhydrocarbon usage, however. This is a shame, as it is a very powerful and easily applied tool.

A cross plot is simply the procedure of plotting one geophysical parameter data set against another on graph paper or mathematically. A single data point gives some information, but it can be ambiguous. The clusters of data points from two or more methods can be very diagnostic. Statistical analytical methods can further refine the process.

Thus, a single density reading of 2.3 g/cm^3 (Figure 9.17) could be in a clay, a shale, a sand, a shaly sand, a very porous limestone, or a gypsum. It could also be an error or a glitch. It would not be a carbonate, unless it were very porous because they are too dense. It could not be a coal nor an evaporite because they are too light. Therefore, a single value can be ambiguous. A cluster of points (Figure 9.18) at 2.30 g/cc, in a 2.5 g/cc background, becomes more informative, but it could still be a shale within a porous limestone or in a cemented sandstone. It can also be used to show trends. If, however, a gamma ray set of readings is plotted against the density readings, it immediately becomes apparent that the zone is a shaley volume within a porous limestone (see Figure 9.19). This is an elementary illustration of the value of cross plotting.

Cross-plotting techniques can easily be programmed in a computer to recognize various minerals, fluids, and situations and put a name to them. This is the background of modern oilfield imagery and log presentation. The method is, of course, an application of solving a set of simultaneous equations. *Introduction to Geophysical Formation Evaluation* and *Standard Methods of Geophysical Formation Evaluation* have many examples of cross plotting. These are a major tool in modern oilfield log interpretation. These methods can easily and profitably be applied to nonhydrocarbon applications.

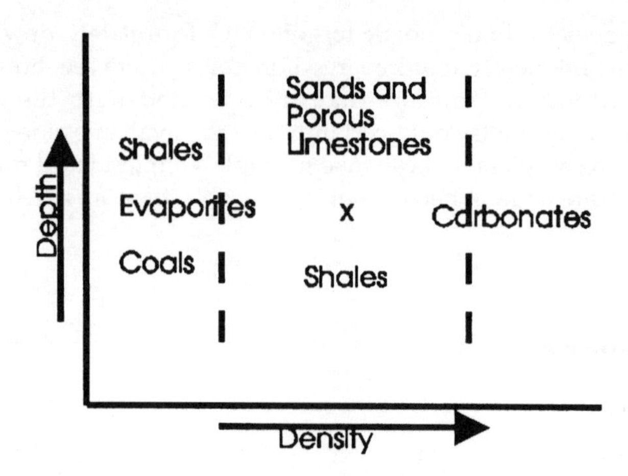

FIGURE 9.17
A single density reading.

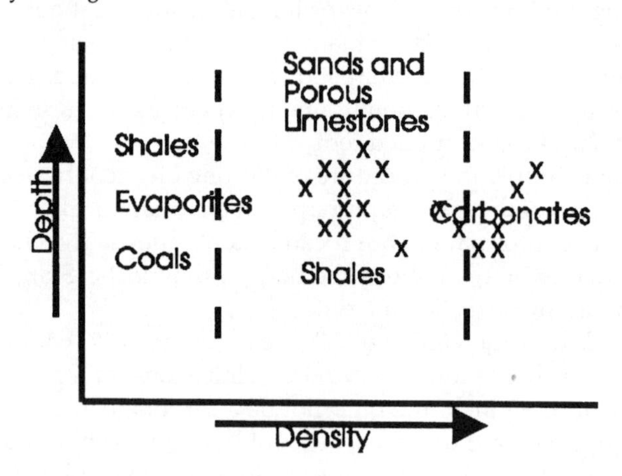

FIGURE 9.18
Multiple density readings.

Cross plotting can separate and identify lithology, porosity, fluids, and/or component rock materials. When quantitatively presented in log form, it can show the amount of these features and their locations.

9.11.2 Data Handling and Transmission

The standard mode of data handling and transmission, in the past, has been analog, and is still widely used. In analog transmission, the raw signal is generally a voltage or current whose amplitude continuously represents some function of the information (for radioactive devices, the signal is generally a series of randomly spaced pulses whose number per unit time is a function of the desired signal).

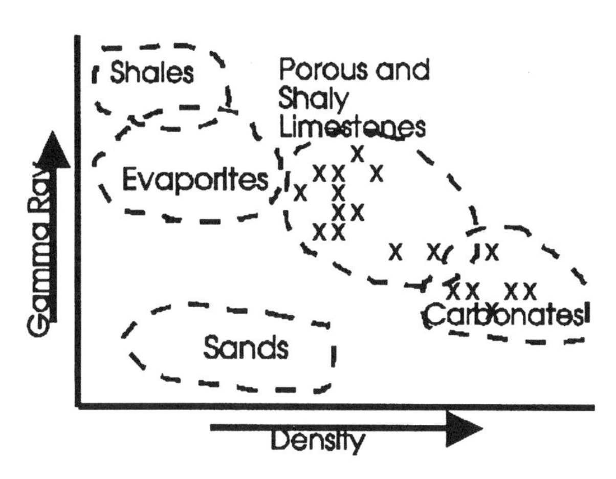

FIGURE 9.19
A two-curve cross plot.

The signal is generally amplified before transmission to overcome transmission losses (line losses). Radioactive pulses are generally squared off and may be divided by a known, constant digit to reduce losses from near-coincidence occurrences. After transmission to the receiver and processor, the signals are usually filtered to remove predictable noise and then amplified. Radioactive pulses are amplified and squared off again. The signal amplitude or the number of pulses per unit time are read and the resulting variable amplitude signal sent to the recording device. The analog recording device may be a line-graph recorder, simple meter, counting circuit, numerical display, alarm circuit, and/or tape recorder.

During the late 1960s, various geophysical companies were investigating digital methods of data handling. Compact, rugged, solid-state computers and other digital devices were beginning to appear on the market. In 1974, Century Geophysical Corporation, of Tulsa, OK, demonstrated and fielded a fleet of fully digital borehole mineral logging units. Their advantages were immediately apparent. Most of the logging contractors converted to digital or hybrid digital/analog systems.

9.12 Digital Signal Transmission

Transducer or detector signals may be either analog or digital. Analog signals are sampled periodically at a rate high enough so that the signal variations may be faithfully reproduced. A simple handling circuit may handle several inputs, sampling them in sequence. The output signals are precise and controlled by a stable clock (oscillator). The samples must be frequent enough to be able to faithfully reproduce the original signal.

At the receiver, the pulses are separated by time and routed to the proper processor circuits. The pulses are then assembled to recreate the original signal. The presence or absence of a pulse is all that is needed. Noise and many of the transmission distortions are simply ignored and disappear. Transmission may be by wire line, radio signal, fiber-optic cable, infrared pulses, or any other feasible method.

9.13 Presentations

The field use of computers has become almost universal. This is true even when the process is analog. This allows the data to be processed rapidly and accurately in the field. Thus, it can be ready for use within minutes of recording.

Because the operation of a modern computer is fast (and getting faster) and memory devices are growing exponentially, it is possible now to offer finished data in the field that would have formerly required hours, days, and even weeks to process. Many of the recent changes in oilfield logging have been in new and graphic presentations.

Presentations may be in the form of line graphs with respect to time or depth or other useful parameters. Data are often corrected automatically for detector distortions and geometry during computer processing. The data can be presented in visual images showing the relative occurrences and interactions. The possibilities appear endless at this time and are affecting both petroleum work and nonhydrocarbon efforts. Further, one cannot only view a finished product in the field, but the same product can be presented in the office in minutes, even on the other side of the world. The same techniques are used in the proliferation of space probes and subsurface ocean exploration. The vehicles and transmission methods vary, but the idea is the same, whether from a logging truck or from Voyager.

9.14 Data Handling and Transmission

The traditional mode of data handling and transmission has been by analog means. In analog transmission, the raw signal is generally a voltage or current whose amplitude continuously represents some function of the information (for radioactive devices, the signal is generally a series of randomly spaced pulses whose occurrence per unit time is a function of the desired signal).

The signal is generally amplified before transmission to overcome transmission losses (line losses). Radioactive pulses are generally squared off and may be divided by a known, constant digit to reduce losses. After transmission to the receiver and processor, the signals are usually filtered to remove predictable noise and then amplified. Radioactive pulses are amplified and squared off again. The signal amplitude or the number of pulses per unit time are read and the resulting signal sent to the recording device. The analog recording device may be a line-graph recorder, a simple meter, a counting circuit, a numerical display, an alarm circuit, and/or a tape recorder.

During the late 1960s, various geophysical companies were investigating digital methods of data handling. Compact, rugged, solid-state computers and other digital devices were beginning to appear on the market. In 1974, Century Geophysical Corporation, of Tulsa, OK, demonstrated and fielded a fleet of fully digital borehole mineral logging units. Their advantages were immediately apparent. These were followed by most of the logging contractors converting to digital or hybrid digital/analog systems.

The advantages of digital equipment, transmission, and data handling are many. Transmission distortions and losses are minimized. Transmission channels can carry a multiplicity of simultaneous signals. The signals can be read directly by the computer. The computer can route, sort, and process signals rapidly. Signal reproduction can be very faithful. The signals can be displayed in an infinite number of ways. Data can be reduced and corrections made automatically. There are virtually no limits to the distance the signals can be faithfully transmitted. Field maintenance is reduced to removing and replacing a faulty circuit board. Instruments tend to be multipurpose, and this often can be done at a lower cost than with analog methods.

10

Transducers and Electrodes

10.1 Introduction

Our geophysical instruments measure the physical and chemical parameters of our environment. We measure the electrical, mechanical, thermal, and chemical aspects of our surroundings or our target and attempt to interpret these values to give us a representative picture of the portion we are investigating. We do this for a very wide variety of reasons, a few of which are briefly discussed in this series.

In order to measure the character of a portion of our environment we have measuring instruments that receive signals, process them, and present them in a variety of ways. In some cases, they introduce a known energy into the environment and show the change of that energy, due to its passage through the target volume. To do this, we must have devices which form the interface between the environment and the instrument circuit. This device must generate or read a signal into or out of the surroundings and it must supply a faithful electrical (usually, but not necessarily always) representation of the parameter it and the rest of the instruments are designed to examine; or it must supply a well-defined bit of energy into the surroundings for the purpose of being affected by one or more of the parameters being investigated. The device which forms the interface between the instrument circuitry and the surroundings is a transducer. The *Random House Dictionary of the English Language* defines a transducer as "a device that receives energy from one system and retransmits it, often in a different form, to another."

In addition to transducers, some of our geophysical instruments use *electrodes.* Electrodes form an interface between the instrument and the environment and can be considered as a special form of transducer. The *Random House Dictionary* defines an electrode as "a conductor, not necessarily metallic, through which a current enters or leaves a conductor of the non-metallic class, as an electrolytic cell, arc generator, vacuum tube, gaseous tube, etc." We usually find electrodes on geophysical equipment to measure resistivity, spontaneous potentials, and related electrical phenomena.

10.2 Electromagnetic Transducers

Geophysical transducers take many forms. In electromagnetic devices (induction logging, surface electromagnetic, and magnetic systems), the transducers are usually coils of wire. If an electrical current flows through a wire, the wire will be surrounded by a magnetic field. Conversely, if a wire moves through a magnetic field, an electrical potential will be evident along the wire and completing the circuit will result in a flow of current. These effects can be reinforced by winding the wire into a coil so that many serial conductors are in close proximity. Figure 10.1 shows a coil transducer and its associated magnetic field.

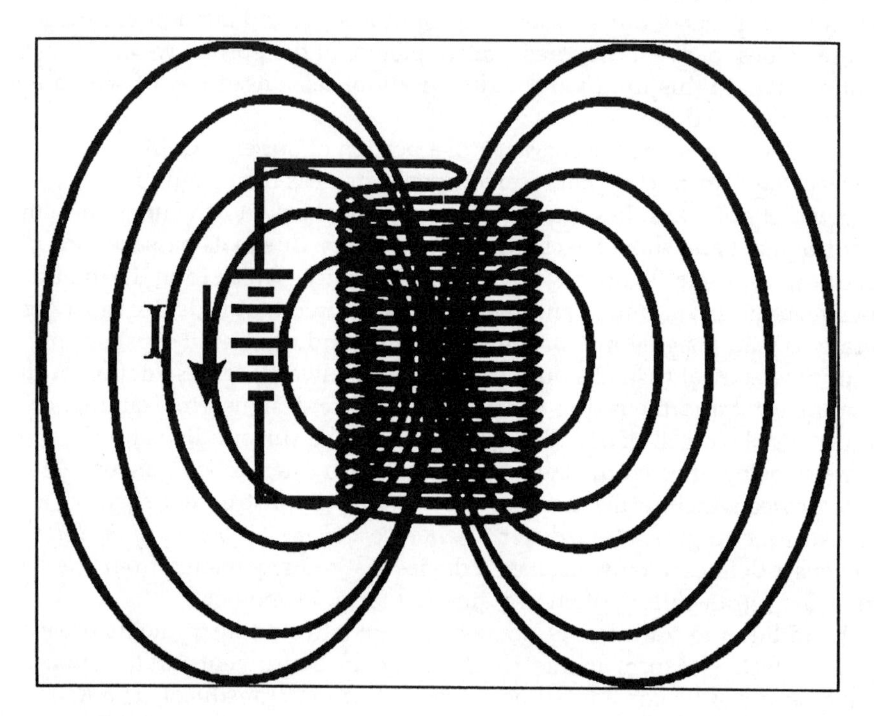

FIGURE 10.1
A coil and associated magnetic field.

Thus, we use coils on induction logging equipment to create artificial moving magnetic fields in the adjacent formation material. Other coils are intersected by the moving magnetic fields after they have been modified by the formation parameters. The alterations of the received fields are mirrored by resulting electrical currents. These, in turn, are interpreted as representing the formation electrical and magnetic properties (electrical conductivity and magnetic susceptibility).

Surface electromagnetic instruments operate in much the same fashion as the induction log. The coils, however, instead of being small enough to fit into a borehole, are many feet or meters (even kilometers) in diameter. Also, the surface magnetic fields are less likely to be focused (or artificially shaped), as are the induction log fields.

10.3 Acoustic Transducers

Acoustic transducers are used to generate and receive mechanical pulses. These commonly take one of several forms. The most common are piezoelectric, but magnetostrictive devices are sometimes used in downhole tools. Surface sources use solid explosives, thumpers, or vibrators. Offshore, the sources may be solid explosives, explosive gases, or collapsing steam from electric sparks.

Piezoelectric devices produce an electric potential across a sensitive material in response to changes of pressure. Likewise, a changing electrical field across the device will cause the device to distort its shape. Thus, downhole acoustic sondes have piezoelectric transducers which respond to electrical pulses from the instrument to put mechanical pulses into the formation material. After a pulse has traveled through the formation material and has been modified by it, the modified pulse will strike a receiver piezoelectric transducer, generating a proportional electrical pulse in the receiving circuit. Figure 10.2 diagrams a piezoelectric transducer.

Magnetostrictive devices can be used alternatively with the piezoelectric units. These transducers use an intense magnetic field from a coil around a ferromagnetic core. The field will cause the core to change shape, resulting in a mechanical pulse into the formation material. Figure 10.3 shows the principle of a magnetostrictive transducer.

10.3.1 Geophones

Geophones ("jugs") are specialized low frequency microphones used to detect mechanical seismic pulses. They are essentially inertially coupled or direct-coupled electromagnetic (or occasionally capacitor) microphones. The armature or the coil is mechanically or inertially coupled to the formation, while the other (coil or armature) is held stationery by its inertia. Downhole geophones usually incorporate a mechanoelectric or hydraulic device (clamp) to couple the device to the wall of the hole. Figure 10.4 shows the principle of a geophone.

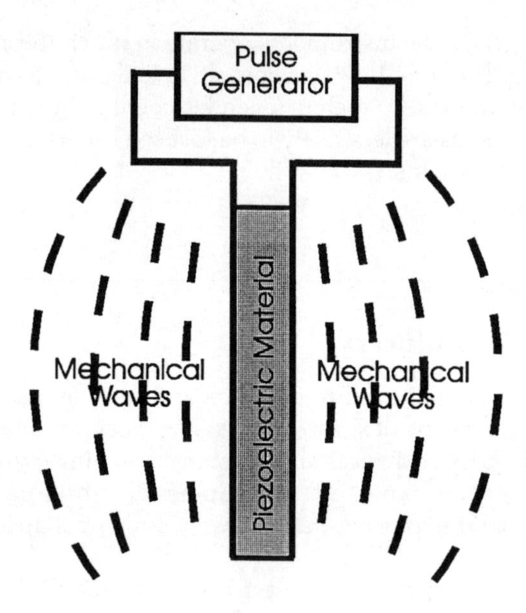

FIGURE 10.2
A piezoelectric cell.

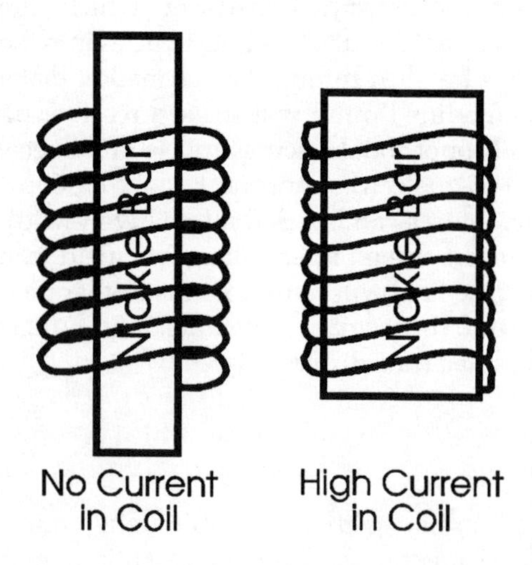

FIGURE 10.3
The magnetostrictive effect.

10.3.2 Seismic/Acoustic Sources

Surface acoustic or mechanical pulses are usually of much greater amplitude than the downhole acoustic tools, simply because of the greater distances and volumes involved. Therefore, different types of pulse (noise) generators are

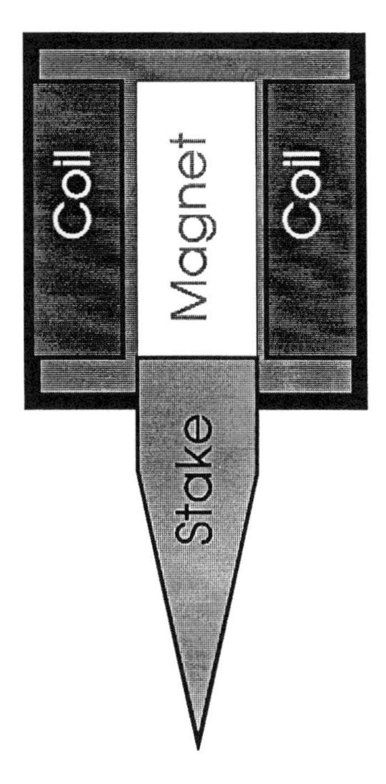

FIGURE 10.4
The principle of the geophone.

used. The traditional source for seismic work is a small explosive charge in a shallow shot-hole. The use of a shot-hole minimizes the surface damage and increases the efficiency of the source. Except for surface shot-holes, explosive sources are seldom used downhole. This is because of the probable damage to a borehole which must be used further. Underwater pulse generators often use explosive mixtures of gases (i.e., methane and air or oxygen) ignited by an electrical spark, but more traditional explosive materials are also used. Some systems use a high current electrical spark discharge from a capacitor storage. The spark rapidly generates steam, which promptly condenses and implodes to generate a pulse. Some surface seismic systems drop a heavy weight to hammer the surface. One small portable system uses a hand-held hammer on a pad on the surface for shallow pulses. These are popularly called "thumpers". A modification of this idea uses an unbalanced flywheel on a surface pad to generate a continuous wave. This is the "Vibroseis" system. Downhole sources for vertical seismic profiling (VSP), cross-hole, and multiple readings are often small variations of the thumper idea or the vibrator. These sources are usually clamped to the side of the borehole. Gas explosions and/or very small solid explosives are occasionally used downhole. They can easily, however, cause hole damage and stuck equipment. Thus, they are not very popular.

10.4 Electrodes

Electrical measurements often require a different form of transducer, the electrode. An electrode is a complex device. A discussion of its properties could well belong in Chapter 9, Section 9.7.1, Redox. The electrode is typically involved in a rather complex electrochemical process which defines the performance of the electrode.

If a piece of base metal is put into a salt solution, it immediately begins to react with the ions of the solution. Atoms of the metal will react to form metal ions in the solution and leave electrons in the metal. Initially, a voltage E_0 will be produced on the metal. This will rapidly die away to zero as the system comes to equilibrium (adapted from Garrels and Christ, 1965). The final voltage will depend upon the type of metal. Finally, the voltage will be the "half-cell" voltage (a hydrogen half-cell is arbitrarily considered as zero or reference potential):

$$Eh = E_0 + \frac{RT}{n\mathcal{F}} \ln Q \tag{10.1}$$

where

E_0 = the initial potential.
R = the gas constant.
T = the absolute temperature.
Q = the ratio of the activities of the ions present.
$\mathcal{F}$ = a constant, the faraday.
n = the number of electrons involved (i.e., 2 in the case of a copper/hydrogen cell).

If two metal electrodes are compared, the voltage read between them will be the difference of their respective half-cell voltages. If two electrodes in solution are connected together through an electrical current source, the equilibrium of each will be altered by the addition of electrons to one and the removal of electrons from the other. The charges will flow through the solution by means of the ions moving from one electrode to the other. Figures 10.5A and B show the operation of electrodes.

If the electrodes are to be used with an alternating current (AC) system (i.e., a resistivity system), the requirements of the electrodes will be quite different from those of the electrodes in a direct current (DC) system (i.e., a spontaneous potential, or SP, circuit). An alternating current (AC) system can use a wide variety of electrode materials, as alternating currents build up and then reverse the surface phenomena, resulting in a neutral balance. Alternating currents seldom exist naturally in the earth. The exception, however, is the

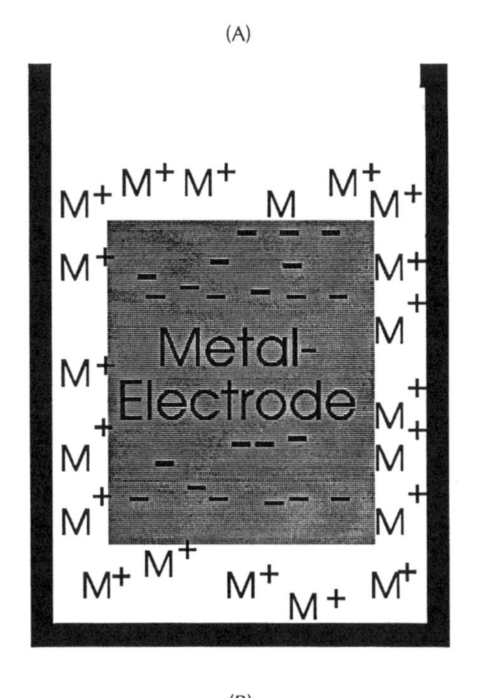

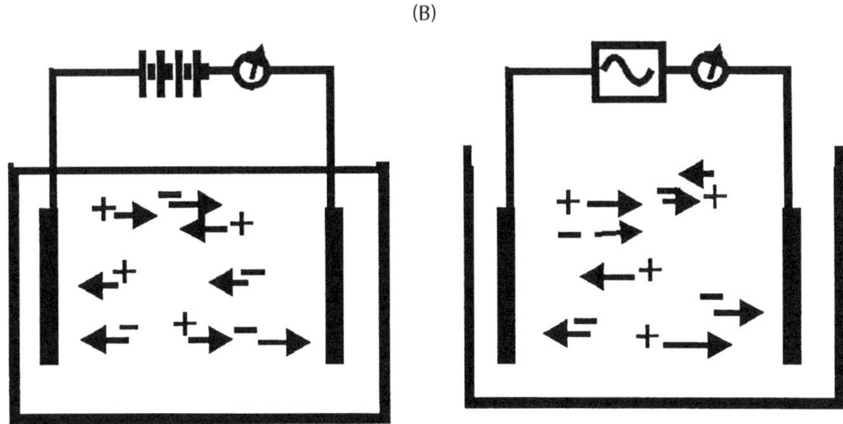

FIGURE 10.5
(A) An electrode in a salt solution. (B) DC and AC electrode action.

wide occurrence of 60 and 50 Hz from commercial power systems. Iron, stainless steel, copper, brass, gold, silver, are all common AC electrode materials.

DC systems such as the low level SP system require careful choice of electrode material. It must be stable thermodynamically, as a change of thermodynamic state can put spurious noise or signals into the system. Also, the two electrodes *must* be of the same material so the half-cell potential of one will cancel that of the other. The chemical activity of the electrode material is important because redox reactions generate electrical energy. The material should be inert, or nearly so (i.e., lead, gold, platinum). The coating of the electrode should cling to the surface and conduct equally in both directions. Lead makes a good electrode material because it is relatively stable thermodynamically and its surface coatings are clinging and nonrectifying. It is inexpensive, relatively rugged, readily available and is easily formed mechanically. Iron and steel make mediocre electrodes because of their thermodynamic characteristics. Their chemical reactivity is against them, also. They are easily fabricated and, because of their large surface area after rusting, they have some uses, especially in AC circuitry. Gold can be used, especially as a thin, noncorrosive coating, in some uses. Aluminum and stainless steel and nickel and cobalt form tight, high resistivity coatings. They are suitable only for higher frequency AC uses.

10.4.1 Half-Cell Electrodes

Electrodes for delicate and/or precise measurements are frequently of the metal-metal chloride type. These are stable and their half-cell voltages are predictable. They are fragile and require constant care. They are difficult to use downhole, although some have been used thusly. Schlumberger Well Services, Inc. experimented with silver/silver chloride (Ag/AgCl) electrodes, a number of years ago, to improve the stability of the SP measurement. The mechanical assembly was difficult and there was not enough improvement in the measurement to warrant the additional problems associated with the half-cell electrodes. Figure 10.6 shows the construction of a half-cell electrode.

10.4.2 pH Electrodes

pH electrodes are modifications of the half-cell type electrode. This measurement is always referred to the hydrogen half-cell potential. In fact, the term *pH* refers to the potential of the hydrogen electrode. The value of the pH is the logarithm of the concentration of the anions in solution:

$$pH = -\log\left[H^+\right] \tag{10.2}$$

The concentration of H^+ is designated as $[H^+]$ and cannot be measured directly. Therefore, an EMF cell (half-cell; a glass electrode) and a reference electrode are placed in the same solution to measure pH. The reference is usually of the mercury/mercuric chloride type ($Hg/Hg_2Cl_2/HCl$, a calomel electrode). See Ives and Janz, 1961. The glass pH electrode is a bulb of a special glass containing an acid solution. This has an inner electrode of a fixed voltage, usually the silver/silver chloride type. The voltage difference between the reference electrode and the pH electrode is proportional to the log of $[H^+]$. The potentials of both electrodes are functions of their absolute temperatures.

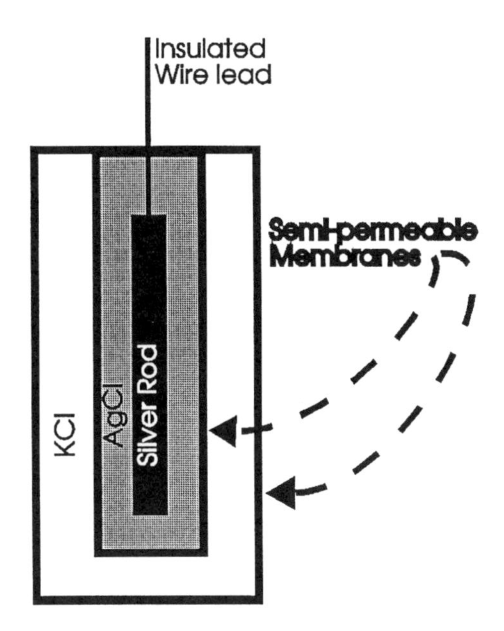

FIGURE 10.6
One type of Ag/AgCl half-cell.

Therefore, a temperature correction must be made if the solution is not at 25°C. A pH of 7 indicates that the concentration of acidic anions equals the concentration of basic anions in the solution. Readings less than 7 indicate varying degrees of acidity. A value of 1 indicates that the solution contains only acid anions. A value above 7 indicates increasing concentrations of basic anions, with 14 being the maximum. pH measurements are frequently plotted against Eh values to diagram mineral stabilities. pH measurements are usually made in the laboratory, although the usual portable pH meter can be used in the field.

This author has made pH measurements on cuttings from a water well near Gillette, WY. The pH changed from slightly above 7.0 in the reduced zone below the top of the water table to below 7.0 in the oxidized zone above the water table. This correlated nicely with the redox deduction from the SP curve and the color changes of the samples.

10.4.3 Specific Ion Electrodes

Specific ion electrodes are a variation of the standard half-cell, and are designed to respond to a specific ion type, such as the sulfate ion ($SO_4^=$). These depend upon a redox reaction with a specific type of ion. The electrodes are usually referred to a hydrogen half-cell or another stable type of half-cell reference electrode. The specific ion electrodes have been in use in the laboratory for many years. Field use, especially downhole, has mainly been experimental.

10.4.4 High Frequency Transducers

Some surface and downhole systems use radiofrequency currents. These systems usually use a dipole or a folded dipole antenna. These often can be covered with an insulating coating for mechanical or a thin coating of gold for chemical protection. One electromagnetic system in northern Canada uses the carrier wave from the U.S. Navy low frequency submarine communication system from a monopole antenna. Its receiver is a large coil, and it makes use of the deep surface penetration of the low frequency field from the transmitter.

10.5 Radioactivity Detectors

Unlike many other geophysical detectors and transducers, radioactivity detectors are only suited for reading radioactivity amounts and energies (and sometimes direction). They cannot be used to generate radioactivity signals. These fall into several classes.

10.5.1 Ionization Chambers

The earliest of the electronic radiation detectors was the ionization chamber. It is still used in some equipment, such as domestic smoke detectors. The ionization chamber consists of a chamber filled with an inert gas, often at atmospheric pressure. A pair of electrodes (i.e., the case and a central wire or rod) set up a moderate electric field. In the absence of any ionizing radiation, such as gamma rays, there are few gas ions present. Thus, the current through the chamber is low. The presence of ionizing radiation ionizes some of the gas molecules and a small current will flow between the electrodes. The amount of the current is a function of the amount of radiation. These devices are slow to react, relatively insensitive, prone to noise, and their signal is low. The signal is DC and difficult to amplify economically. They are, however, simple, reliable, and easy to build and maintain. Figure 10.7 shows the principle of the ionization chamber.

10.5.2 Geiger-Müller Detectors

Geiger-Müller (GM) detectors were the next generation of electronic detectors. They helped usher in the radioactivity research, as well as the independent prospector for uranium.

The GM detector is a conductive, rigid cylinder cathode with a central, coaxial, anode wire or rod. It has an intense electric field across it. A radioactive particle or photon will dislodge electrons from the outer cathode. These

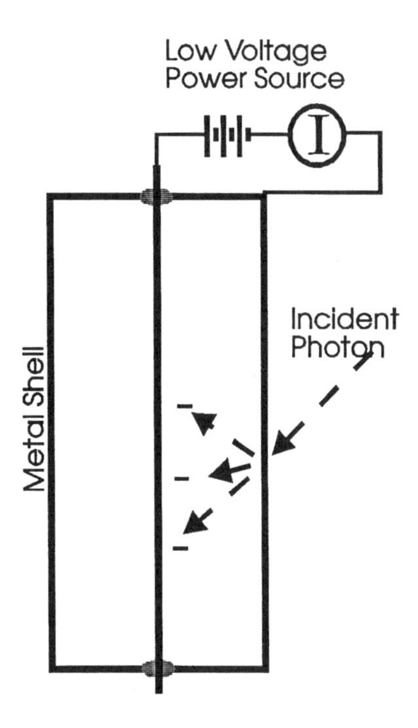

FIGURE 10.7
An ionization chamber.

electrons will accelerate rapidly toward the central anode, ionizing the gas molecules on the way. The gas ions will also be accelerated rapidly enough to further ionize the gas. The result is a conductive cascade which will cause a large current to flow. A combination of the proper gas mixture and external quenching circuit will stop the current flow, once it is well started. This results in a large pulse for each event detected. Little or no amplifying circuit is needed. The pulse length will depend upon the design of the tube, but will be from 100 to 500 μsec long. Figure 10.8 shows the construction of a GM detector.

The GM detector has a high signal level, simplifying the external circuitry. The power supply need not be especially stable, as long as it remains above the level necessary for the GM action. The pulse height will depend upon it. It is not sensitive to the energy level of the radioactive particle, and its pulses are long. Thus, it is unsuited for high level or spectrographic work, but it is a rugged, simple, maintenance-free piece of field equipment.

10.5.3 Proportional Counters

The proportional counter lies between the ionization chamber and the GM detector, in operation. Its internal operation is similar to that of the GM detector, except the voltage field is not high enough to cause a cascade. Therefore,

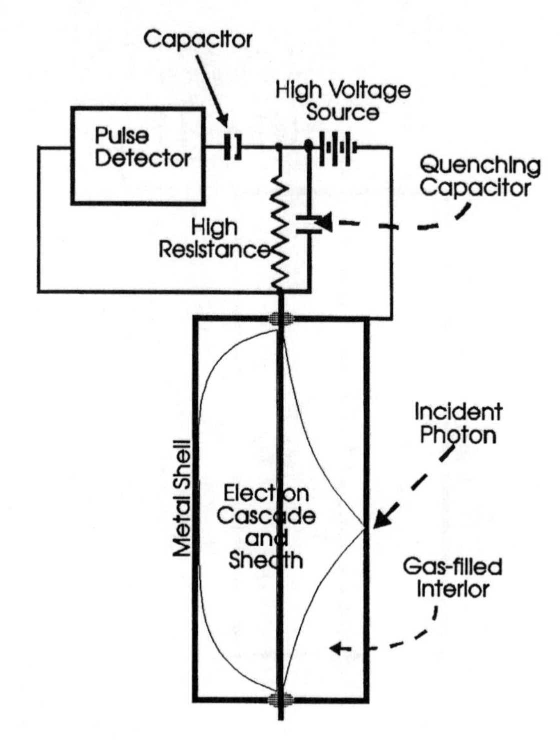

FIGURE 10.8
The principle of the GM detector.

the ionized path of a radioactive event will drift to the anode and result in a pulse whose amplitude is a function of the energy of the particle causing the ionization. This can be used for spectroscopy, but requires a very stable power source and external amplification. It is temperature sensitive. To be representative of the true energy of the radioactive event, the energy of the particle must be fully expended in its flight within the detector. Because the sensing material is a gas, this is difficult to achieve.

Proportional counters are not much used, because of their drawbacks. They are bulky and have an uncertain zero point. The only one in common use is the helium-3 neutron detector in its various forms.

10.5.4 Scintillation Detectors

Scintillation detectors are almost universally used in modern geophysical gamma ray equipment. Some neutron equipment, also, uses scintillation detectors. Excellent prepackaged gamma ray detectors have eliminated most of the maintenance problems that used to plague gamma ray systems. This equipment, for the most part, is now rugged, sensitive, and relatively trouble-free.

The heart of the scintillation detector is a crystal which will fluoresce or scintillate with the passage of an ionizing radioactive particle through it. The passage of the particle leaves a path of light, as the particle expends its energy in the relatively dense crystal material. If the particle expends all of its energy within the crystal, the length of the path, and thus the amount of light, are proportional to the energy of the particle. The crystal must be as dense and as transparent as possible.

The amount of light is minute. In a NaI-Tl crystal, it is about 2 to 4 μsec long. It is detected by an external photosensitive surface of a photomultiplier (PM). The photosensitive surface is a low work-function material (usually a selenium compound) on the inside of the optical glass end of the PM tube. As the light hits the photosensitive surface, a few electrons are emitted. The number will be a function of the amount of light incident upon it. These electrons are guided and accelerated by an electric field to an intermediate anode, a "dynode". Each dynode has a low work-function surface. At the first dynode, they hit and eject more electrons, which are accelerated and guided to a second dynode, and so on. By the time the electrons have reached the final anode, they have been multiplied by a factor of 10^5 to 10^7, depending upon the tube design. At this point, the burst of electrons forms a pulse of several millivolts and about 1 μsec or less long. The electrical pulse is then routed through an electronic preamplifier at the beginning of the tool circuitry. Figure 10.9 shows the construction of a downhole logging-type scintillation detector.

In general, the size of the scintillation crystal determines its worth as a spectrographic device, and the sensitivity is proportional to the cross-sectional area presented to the path of the radioactive particles, that normal to the photocell. Thus, a typical downhole mineral crystal will be about 1 in. (2.5 cm) in diameter and 8 in. (20 cm) long. Its cross-section would be 8 in.^{2} (50 cm^2). Ideally, a crystal would be as dense and as transparent as possible, particularly for spectrographic work. The downhole tool geometry places a limit upon the diameter and the transparency upon the length. Laboratory, airborne, and surface equipment scintillation detectors are not usually restricted in size and dimensions as are downhole units. Therefore, the latter are larger and have diameters about equal to their lengths. A length of about 8 in. (20 cm) seems to be the longest practical length for downhole sodium iodide detectors, because of the imperfect transparency of the crystal material. In reality, there is no practical limit to the size of the crystal. Some very large volume crystals get around these problems by using multiple PMs.

Crystals cover a wide range of materials. Their requirements are transparency and fluorescent ability. For many years, the most common gamma ray crystal material has been thallium-activated sodium iodide (NaI-Tl). This material is fairly dense ($\rho_{NaI} = 3.667$ g/cc). Its efficiency and transparency are reasonable. It will deteriorate with exposure to moisture and must be hermetically sealed. Iodine is released when the crystal deteriorates and discolors it, reducing its transparency. It and the PM, of course, must be shielded from visible light. There are other scintillation materials which are used. Some

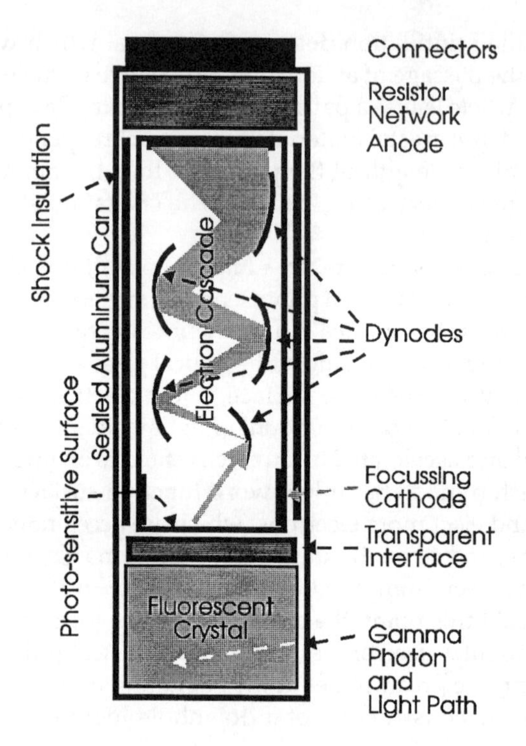

FIGURE 10.9
The construction of a sealed scintillation detector.

have a higher density, which more than compensates for a lower efficiency. Plastic crystals are popular because they do not need the hermetic seal and are not as fragile mechanically. Please refer to *Introduction to Geophysical Formation Evaluation.*

Scintillation detector advantages are a proportional sensitivity to particle energy, small detector size, high efficiency, and short pulse length. On the other hand, they have very low signal outputs and require a very stable high voltage supply and stable amplification. They are somewhat temperature sensitive (which is seldom taken into account in downhole equipment).

10.5.5 Solid-State Detectors

Solid-state detectors (i.e., ultrapure or intrinsic germanium) can be adapted to radiation detection. The passage of a radioactive or other ionizing particle will show up as a pulse of electrons and/or holes proportional to the path length of the particle (i.e., its energy). Because of the density of the material and the high resolution, these devices make excellent radiation detectors. They have very low signal-to-noise ratios and must be operated at cryogenic temperatures to reduce their inherent thermal shot noise. Boiling liquid nitrogen and/or frozen propane is used to achieve these environments. Their size is small and their spectral resolution is about 50 times better than NaI-Tl.

10.6 Mechanical Considerations

The shape, mounting exposure, and material type of an electrode or transducer is very important. Electrodes for downhole resistivity systems must be completely electrically isolated from any conductive material which might bypass the desired path of the electrical current through the formation material. This might appear obvious. However, resistivity downhole sondes have been sold which have conductive paths between electrodes through the sonde body. Surface resistivity and induced polarization (IP) systems are quite sensitive to conductive fence lines and corrosion electrodes.

SP systems must be completely isolated, electrically, from the cable and metal portions of the sonde body. Exposure of a cable head and the zinc-coated cable can result in electrical currents which will superimpose a DC lateral resistivity signal on the SP signal.

It is important that both electrodes and transducers be recessed on the sonde body of downhole equipment to prevent them from rubbing the wall of the hole. The removal and subsequent regrowth of the electrode coatings will put electrical noise into the system. Pebbles and cuttings rubbing against an acoustic transducer will put noise into the system and mechanically damage the transducer.

Mud resistivity meters are susceptible to clogging by thick mudcake. Passages and orifices must be designed to facilitate the mud flow and be self-cleaning. Surface mud resistivity meters *must* be carefully washed with fresh water before and after each use. The sample size, in these instruments, is so small that contamination from the previous measurement is likely.

11

Shielding and Calibration

11.1 Shielding

Shielding of radioactivity addresses two needs. It is used, in conjunction with sources of energy, to protect those using the radioactive material and others in the vicinity. It is also used to control and direct the energy. Both aspects are important in geophysical work. They become more important as techniques new to geophysics are tried and adopted.

11.2 Gamma Radiation

Gamma radiation is electromagnetic radiation, as are visible light, radio signals, infrared radiation, heat, ultraviolet (UV) radiation, and X-rays. All are energy transmission. The main difference is in their individual wavelengths or frequencies. Frequency is an indicator of the energy content. Like all types of electromagnetic radiation, gamma radiation has characteristics of wave motion. These are especially noticeable in the bulkier manifestations. It also shows characteristics of particles. This is especially evident in high energy, low level forms. At the levels at which we will examine the use and handling of the sources of this radiation, the particle nature predominates. Note that with energies of visible light and higher — UV, X-rays, gamma rays — both wave and particle modes are used in describing them. At lower energies — infrared, radio frequencies — the wave mode is used almost exclusively. Figure 11.1 shows the electromagnetic spectrum.

The particles of radiation are called photons. One might think of them as irreducible bundles of wave energy. All photons travel at the speed of light in the medium through which they are traveling. Photons obey the criteria for particle physics. They can be scattered, they have a momentum of $h\nu/c$, they have no electrical charge, no magnetic moment, they have a spin moment, and their energy is proportional to their frequency. A photon particle is small enough that it can, in many of our purposes, be treated as a point. Forces acting upon it are concurrent and it is affected by such forces as gravity. On the

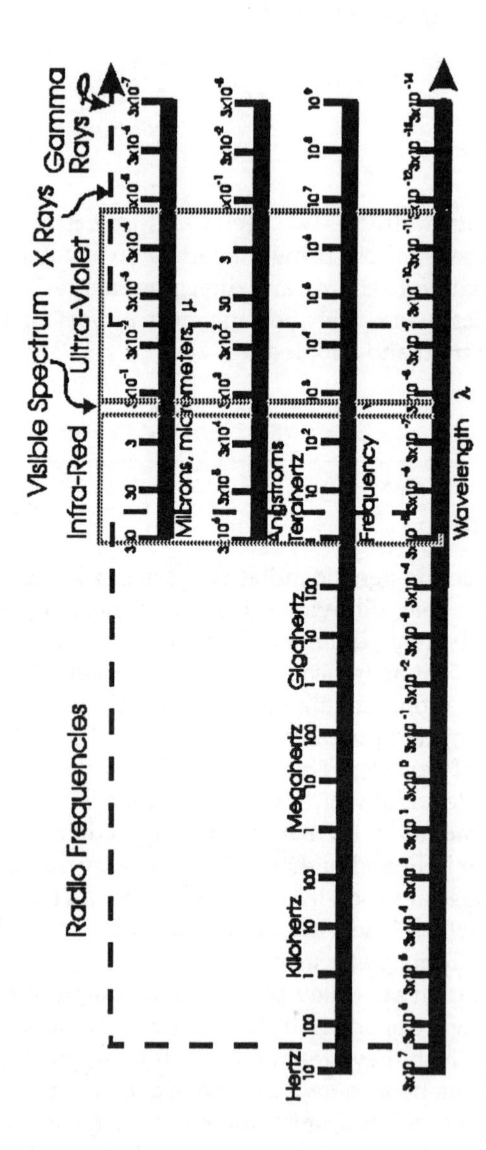

FIGURE 11.1
The electromagnetic spectrum.

other hand, photons have the characteristics of waves. A photon has a wavelength (ν), it travels at the speed of light (c), it can be refracted, and it shows interference phenomena. Both modes represent energy. The energy content is proportional to the wavelength: E = hν.

As a photon travels through a detector (or any other material), it dissipates its energy by interacting with (scattering from) the nearby orbital atomic electrons. It leaves an ionized path, which contains its lost energy. In a scintillation detector crystal, the ionized path is a tiny streak of light, which is the energy from the ionized atoms, as they return to their normal state. If the detector is a gas-filled tube, such as a Geiger-Müeller (GM) detector, The ionized trail forms a conductive path of ionized gas molecules or atoms. This allows the flow of a current in an ionization chamber.

If the photon dissipates all its energy within the detector (scintillation and proportional detectors), the energy of the photon can be determined. The photon, however, will cease to exist. If it does not lose all of its energy within the detector, it will continue on its passage, but at a lower energy. The actual path within the detector will continue for a period of time after the passage of the photon. The decay time will be on the order of picoseconds to milliseconds (or longer), depending upon the detector type. During the decay time, the ionization will decay exponentially. The result is a characteristic shape for a decay pulse, such as shown in Figure 11.2.

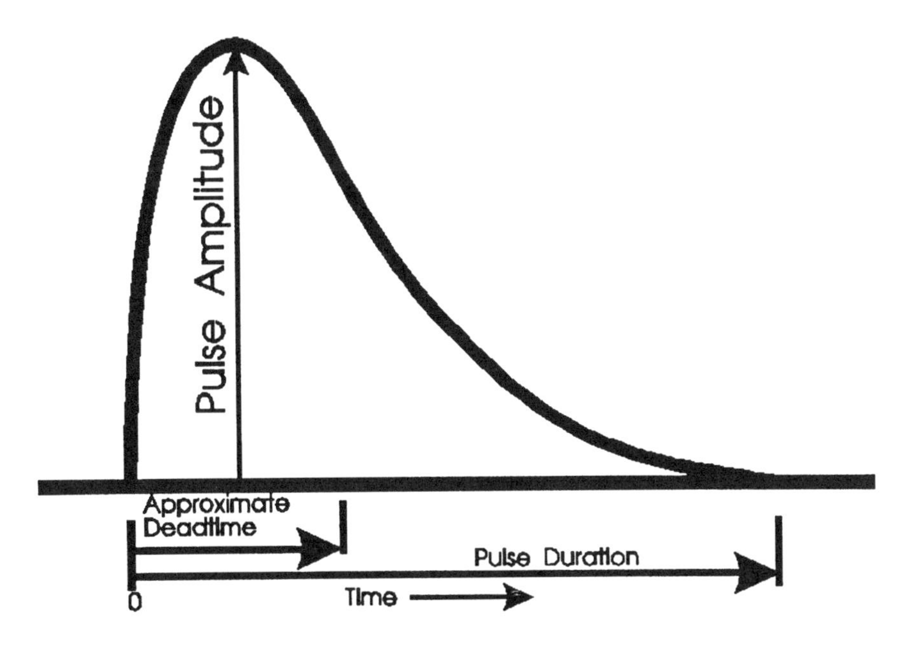

FIGURE 11.2
The usual electronic pulse shape.

The denser the detector active material, the more likely it is that the photon energy will be completely expended within the detector. Therefore, detector materials are generally made as dense as possible.

The interaction of the photons within the detector material (or any other material, for that matter) are classed as Compton (elastic) scattering, photoelectric absorption (inelastic scattering), and pair production. Refer also to *Introduction to Geophysical Formation Evaluation.*

11.2.1 Compton Scattering

Compton scattering is an elastic process. The absorption coefficient associated with it is usually shown as "σ". In this interaction, the photon interacts with an orbital atomic electron. The collision imparts part of the photon energy to the electron in billiard ball fashion. The photon will continue on on a different path and with a lower energy (a longer wavelength, λ). The expression describing this is

$$\lambda_1 - \lambda_0 = \frac{h}{m_0 c}(1 - \cos\theta) \tag{11.1}$$

where

λ_0 = the original wavelength of the photon.
λ_1 = the photon wavelength after collision.
h = Planck's Constant, 6.626176×10^{-27} erg sec.
m_0 = the rest mass of the electron with which the photon collides.
c = the speed of light in that medium.
θ = the angle of departure of the photon from its original path.

The electron will recoil at an angle ø, to the photon original path, with an energy E_e,

$$E_e = \frac{2m_0c^2 E_\gamma^2 \cos^2\phi}{\left(E_\gamma + m_0c^2\right)^{2-} E_\gamma^2 \cos^2\phi} \tag{11.2}$$

where

E_0 = the original photon energy.
ϕ = the recoil angle of the electron, to the original photon path.

The Compton apparent cross section (apparent target size), C, to the photon, is

$$C = \frac{\sigma A}{0.6025} barns \tag{11.3}$$

where

σ = the Compton attenuation or absorption coefficient.
A = the atomic weight of the incident atom.
Z = the atomic number of the incident atom.

11.2.2 Photoelectric Absorption

Photoelectric absorption is the inelastic interaction of a photon with an atomic electron. The photoelectric attenuation coefficient is τ. In this reaction, the photon does not have enough energy to rebound at a lower energy after the collision. The photon energy is completely absorbed by the electron and ceases to exist. The electron is ejected from its orbit (a photoelectron) and will have the energy, E_e:

$$E_e = E_\gamma - E_{binding} \quad (11.4)$$

where $E_{binding}$ = the binding energy of the atomic orbital electron.

The atom is left in an excited state, with the energy E_B. It will shortly emit the energy in the form of characteristic wavelength X-rays or visible light or Auger electrons. The cross section, P, of the atom, in the photoelectric reaction, is

$$P = QZ^5 n \quad (11.5)$$

where $n = m_0c^2/h\nu_0$

The absorption coefficient of the photoelectric reaction, τ, is

$$\tau = 3.08 \times 10^{-9} \frac{P}{Am_0c} \quad (11.6)$$

where P = the photoelectric cross-section per atom.

11.2.3 Pair Production

In the pair production reaction, if the gamma photon has an energy of 1.022 MeV or more, it may produce a photoelectron and a positron, each with an energy of 0.511 MeV. The photon is annihilated in the process. The probability of this reaction occurring is proportional to Z^2 of the incident atom and increases rapidly with the quantum energy, above the threshold. The probability will level off at high energies. Almost immediately, the positron will react with any nearby electron with their mutual annihilation and the creation of two photons of 0.511MeV energy, each. The energy of the pair is E_0 – 1.022 MeV. Pair production occurs in the field of the atomic nucleus, but does not involve a nuclear reaction. There must be another body present to preserve both the energy and the momentum. The pair production attenuation factor, k, is

$$k = k_{pb}\frac{207.2}{A}\left(\frac{Z}{82}\right)^2 = 0.0308k_{pb}\frac{Z^2}{A} \tag{11.7}$$

The values of k_{Pb} are shown in Table 11.1.

TABLE 11.1

Values of the Attenuation Coefficient of Lead, k_{Pb}

E_0/m_0c^2	E_0 (MeV)	k_{Pb} (cm^2/g)
2	1.02	0.0010
3	1.53	0.0037
4	2.04	0.0102
5	3.06	0.0225
10	5.17	0.0320
15	7.66	0.0407
20	10.22	0.0407
33.3	17.00	0.0555

Adapted from White.

11.2.4 Total Attenuation Coefficient

The total attenuation coefficient, μ, then, for any material, is

$$\mu = \sigma + \tau + k \tag{11.8}$$

This coefficient represents only the interaction of a gamma photon with the atomic orbital electrons. Any nuclear interactions are negligible.

11.2.5 Shield Thickness

When the attenuation coefficient of the type of shielding material is known, it is possible to design any gamma ray carrying shield, storage vault, personnel shield, or collimator. The amount of radiation after shielding is simply the initial amount minus the blocked or attenuated amount:

$$Q_\gamma = Q_{\gamma 0} - Q_{\gamma B} \tag{11.9}$$

Q_γ must conform to design goals. $Q_{\gamma 0}$ will be known or can be measured and $Q_{\gamma B}$ is the attenuated (calculated) amount.

11.2.6 Collimation

Though gamma rays resemble visible light in that they are both electromagnetic radiation and both exhibit wave and particle action, the shorter wavelengths of the gamma radiation cause them to exhibit some differences from visible light. Gamma rays are much more energetic than visible light rays. Visible light can be reflected from a mirrored metallic surface and refracted with a glass lens. Neither is possible with X- or gamma rays. The problem is, of course, that the visible light rays are long, compared to the diameter of the atoms of the reflector, while X- and gamma rays are nearer to and shorter than the atomic diameter. Therefore, other methods must be used to create and direct a beam of the shorter waves.

Total reflection from the surfaces of crystals can be used to reflect some of the shorter wavelengths. This is taken advantage of in X-ray crystallography.

Some of the radioactivity instruments, whether downhole, surface, laboratory, or airborne, require a direct beam of gamma rays, either entering or leaving. This is achieved by shielding the source (or detector or both) in the same way as designing a storage or personnel safety shield, but leaving a small opening to allow the unimpeded passage of the gamma rays, X-rays, or other high energy particles. The downhole gamma ray density tool is a good example. It has a collimated beam from the source of gamma rays and a collimated opening to the detector. Both are sealed with windows of aluminum. The system must be shielded from the borehole, at the back of the tool. Figure 11.3 illustrates the principle of collimation.

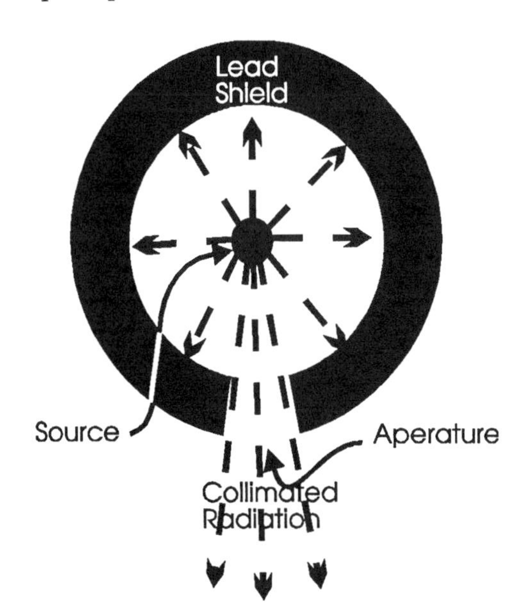

FIGURE 11.3
The principle of collimation.

Massive, high density metals are excellent for collimators and shields. Lead is often used because of its relatively high density, availability, and low cost. Tungsten is a better material, however, in spite of its higher cost. Lead has a density of 11.3 g/cm^3. Tungsten has a density of 19.35 g/cm^3 or 1.7 times greater. This makes tungsten about 4.3 time more effective than lead as a shield or collimator.

11.2.7 Shielding Determinations

The mass attenuation coefficient, μ_m, is a measure of the scattering and energy absorption of a material for electromagnetic radiation. The several elements of this coefficient are shown in Equation 11.8. The linear mass absorption coefficient is the mass absorption coefficient times the density of the material:

$$\mu_l = \rho \mu_m \tag{11.10}$$

The general absorption relationship for any situation is

$$I = I_0 e^{-\mu_l \chi} \tag{11.11}$$

where

I = the final intensity of radiation.
I_0 = the original intensity of radiation.
μ_l = the linear absorption coefficient.
χ = the thickness of the shielding material, in centimeters.

Using Equation 11.11, any shield or collimation thickness can be calculated. The half-value thickness, "$\chi_{1/2}$", is also a convenient way to estimate the same values. Each thickness of $\chi_{1/2}$ reduces the radiation by 50%. The half-value thickness is:

$$\chi_{1/2} = \frac{0.693}{\mu_l} \tag{11.12}$$

11.3 Neutron Source Shielding

Neutron source shielding and collimation must depend upon the mechanical dissipation of the energy of the neutron by direct collision with a hydrogen atom. This slowing down of the neutron (moderation) is usually accomplished by allowing the fast neutrons to collide with hydrogen nuclei (protons). Because neutrons and protons have almost the same masses, the transfer of energy is maximum (up to 50% for each collision). Moderation with hydrogen requires, on the average, only 12 collisions with a hydrogen atom to reach thermal energy. Neutron shields, therefore, consist mainly of hydrogenous materials, such as paraffin, polyethylene, organic oils, or water. Frequently a high cross-section material, such as a boron or cadmium compound, is added to the paraffin.

After the neutrons have moderated to a thermal energy, they are easily captured by many materials. The capture gamma rays can be shielded by a sheath of lead on the outside of the shielding or collimation housing.

Collimation of neutrons, of course, will follow the same patterns, but with an orifice for the entrance or exit of the beam of high energy neutrons. The orifice window may be almost any material which does not contain hydrogen or other high cross-section elements.

The amount of paraffin needed to shield a neutron logging source is fairly extensive. Figure 11.4 shows the result of an early experiment. In this experiment (Cork, 1957), a source of fast neutrons (Radium-Beryllium — RaBe) was placed at a fixed distance from a neutron detector. Paraffin was used as a shielding material around the source. The thickness of the shield was varied from 0 to 24 cm (0 to 9.4 in.). With a thickness of 10 cm (4 in.), the number of fast neutrons had dropped to 40% of its unshielded amount. The number of thermal neutrons, however, had risen to 833% of its original amount. With 24 cm (9.4 in.) the number of fast neutrons was very low, but the number of thermal neutrons was still 140% of the source amount.

The moderation in the paraffin, of the fast neutrons from the source, resulted in the number of thermal neutrons increasing through shielding thicknesses up to 10 centimeters (4 in.). It thus appears that too little shielding material can be worse than none or too much. Figure 11.5, however, shows the change of the amount of neutron scattering as a function of the energy of the neutron. It appears from this that about 30 cm (12 in.) of paraffin would have lowered the number of thermal neutrons to about 10% of the original source of thermal emanation and virtually eliminate the fast neutrons.

To reduce the thickness of paraffin or any other hydrogenous material needed to exceed the peak thermal neutron generation thickness, a high cross-section material such as boron or cadmium is frequently added to the hydrogenous material in powdered, granular, or sheet form.

11.3.1 Source Storage

Locked storage of the gamma and neutron sources, when they are not being used, is absolutely necessary. In fact, it is required by law in most places. The location should be a convenient one, where outsiders cannot easily gain access. The easiest type is a steel-lined concrete pit below ground. The steel lining is often omitted, but is fairly easy to include. The source should be left in its portable shield and lowered with a hooked rod. A block of plastic or

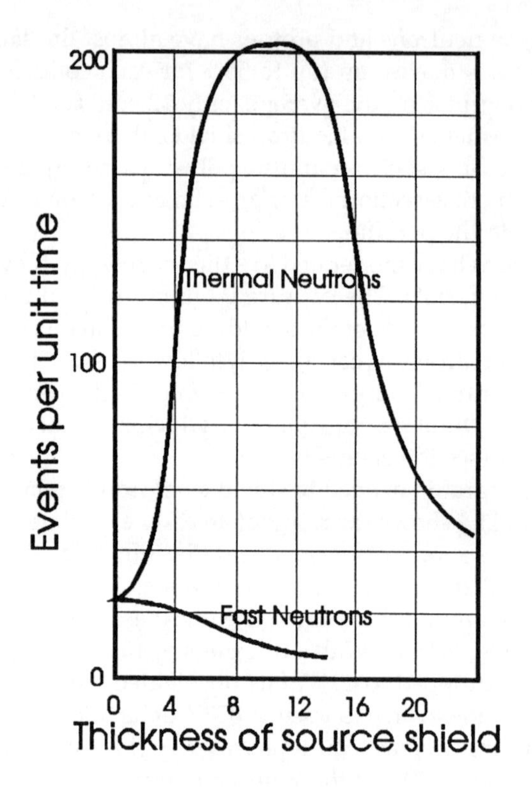

FIGURE 11.4
The moderation of fast neutrons by paraffin.

paraffin can be added on top of the source container, for additional long term safety. A steel cover and lock must be used at the entrance. Figure 11.6 outlines such a storage facility.

If a below-ground storage facility is not practicable, a shed-type cabinet can be located away from heavy traffic areas. The shed should be made of concrete and have locked steel doors. Inner doors of heavy plastic can aid shielding.

11.4 Calibration

The regular calibration of *all* geophysical instrumentation is important. There are enough variables and uncertainties in our work, without adding more that can be avoided. Also, *all* components vary with time and temperature. No instrument is stable. Some are just more stable than others; but all vary in response with time, temperature, pressure, and many other factors. Calibration should be a fundamental part of any new instrument design, new development, production, and a routine part of maintenance. A regular schedule

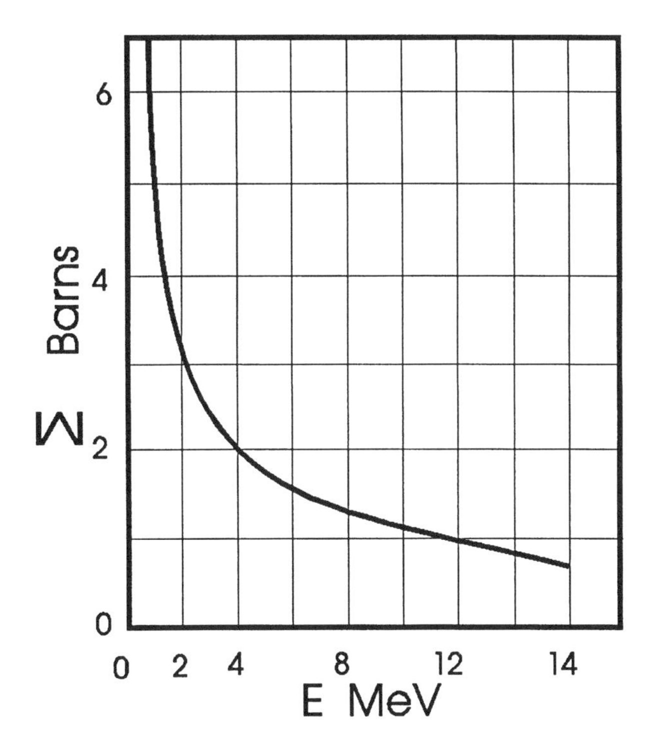

FIGURE 11.5
The cross-section of hydrogen, as a function of neutron energy.

should be set up for routinely calibrating field equipment. Field devices should be supplied and used for checking before and after each day of measurements. In some cases it is feasible to field calibrate before and after each measurement. Computer operations can be programmed to automatically perform calibrations at proper intervals or require calibration before an operation.

Calibrations can be divided into several categories, each with its own urgency:

1. They should be performed as a normal part of the development of every instrument.
2. A final, basic, absolute calibration should be performed before any new instrument, new design, modification, or a new production batch is released. It is not good enough to assume that, because each instrument has identical circuit elements, it will respond in the same way as any other of the batch. Components vary, sometimes widely.
3. If an instrument acts suspiciously or has been damaged, it must be recalibrated.

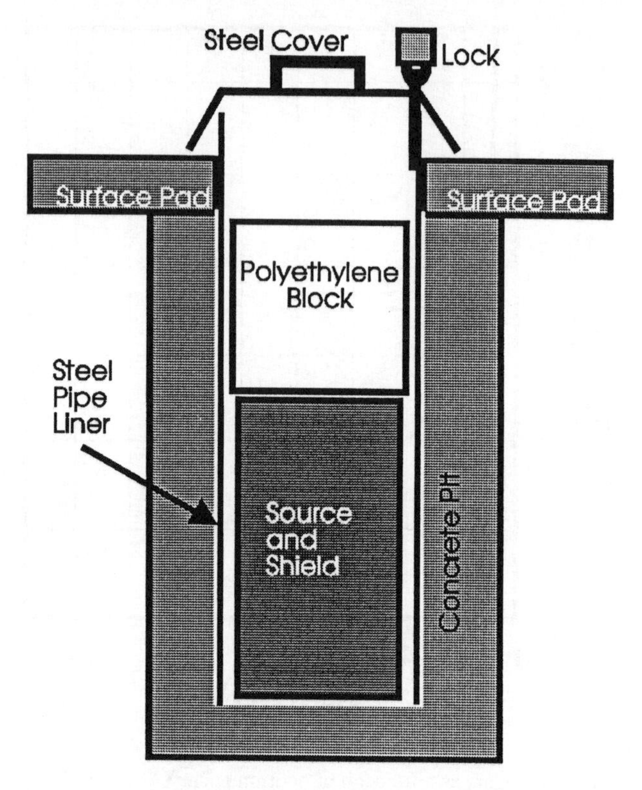

FIGURE 11.6
A typical neutron source storage pit.

4. The individual instruments should be checked against a good standard on a regular schedule.
5. If there is any question about the calibration, it should be thoroughly checked before a project begins.
6. A field calibration each project day is not unreasonable.
7. It is not unreasonable to calibrate some instruments before each measurement.

11.4.1 Resistivity and Conductivity Equipment

Resistivity and conductivity equipment are usually assumed to be very stable. They are often very good, but they are not stable. A good instrument design will be "insensitive to supply voltage". Voltage regulators have degrees of and limits to their control. Fresh batteries are no guarantee of stability, as their output and capacity depend highly upon their internal temperature.

Resistivity (and some conductivity) equipment often relies upon a precision resistor for field calibration. This resistor must also be temperature stable, which is not always the case. Precision resistors usually have a ±1% tolerance, which is adequate for most field work. Laboratory work may demand better. Figure 11.7 shows the principle of resistivity circuit calibration. Figure 11.8 shows one method of field calibrating induction logging equipment.

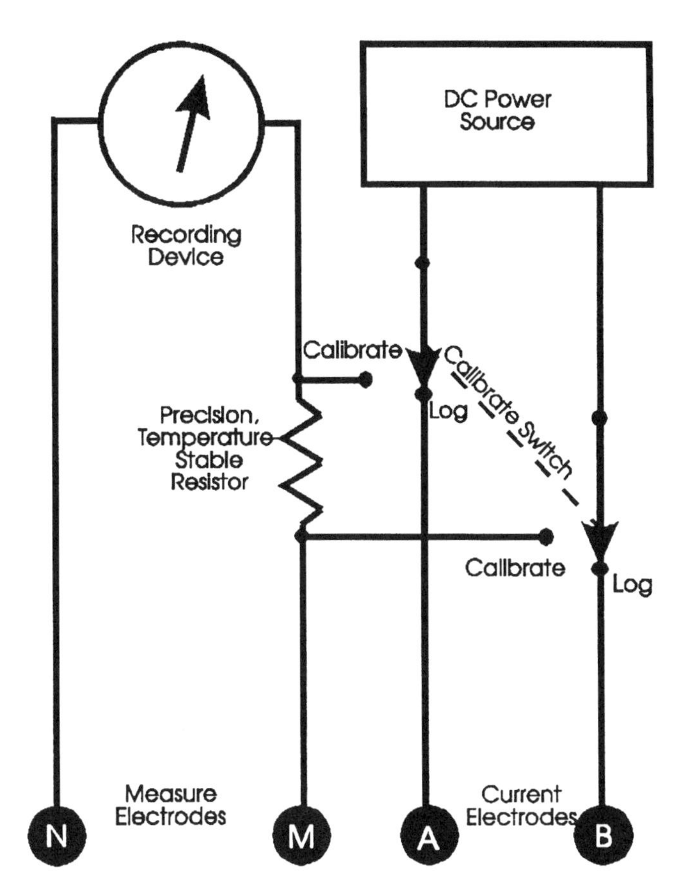

FIGURE 11.7
The principle of resistivity field calibration.

During development, the new resistivity or conductivity instrument should be checked in a large-enough body of water whose resistivity has been carefully determined. This is surprisingly easy to do. A tank, whose dimensions are large enough, can be adjusted to any resistivity by the addition of precise amounts of rock salt. A good, high capacity circulating pump can ensure thorough mixing and distribution. An instrument that has a large or deep volume of investigation may need to be checked in a lake, river, or an ocean.

A precision resistance calibration tells nothing about the geometry of measurement of the device. This must be determined during the design of the

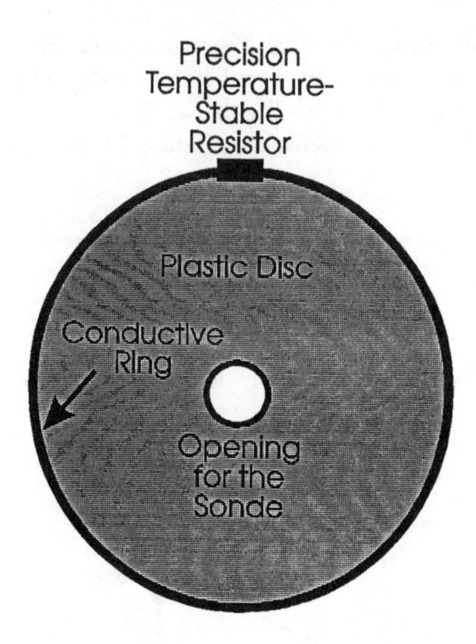

FIGURE 11.8
The field calibrator for the induction log.

instrument. Because the measurement geometry depends mostly upon the mechanical and electrical fixed formats of the probe, it can usually be safely assumed that each identical tool will have the same measurement geometry in an identical environment. There have been some resistor networks that were designed to check geometry. These, however, can become very complex. Computer modeling can give good results, especially during the preliminary design stages. This, latter, should be checked with the actual instruments, under actual field conditions. A large or deep investigation can often be checked with a small, exact-scale model. The field around the device can be checked with a small probe.

11.4.2 Acoustic Equipment

Acoustic equipment calibration has been difficult with respect to relating readings precisely to geological parameters. This was due to the large number of unknown responses of geological features. Most of these problems have been resolved, at this time. Good identification of the problems and solutions have come about by comparing the responses of the downhole tools with surface seismic and vertical seismic profile (VSP) responses.

11.4.2.1 Seismic and VSP

Seismic calibrations have relied greatly upon referring to responses in well-known conditions. These have been from well-documented oilfields, large bodies of water, well-known formations, and core analyses. The information from laboratory measurements of samples was fed into seismic analyses. In many cases, these values remained uncertain because of the inability to duplicate deep-earth conditions. Wide use of acoustic well logging systems and their associated studies furnished good *in situ* information. The combinations of seismic, VSP, and acoustic logging information has resulted in firm knowledge of the responses of all three systems.

Acoustic logging equipment calibration is often accomplished in carefully constructed models. This author was involved in the construction of one such model for a major oilfield logging contractor. This model was designed for acoustic equipment, but served well for several other types of porosity measuring systems, especially neutron porosity systems. Figure 11.9 shows a cross section of one such model. Also refer to Belknap, et al., 1959.

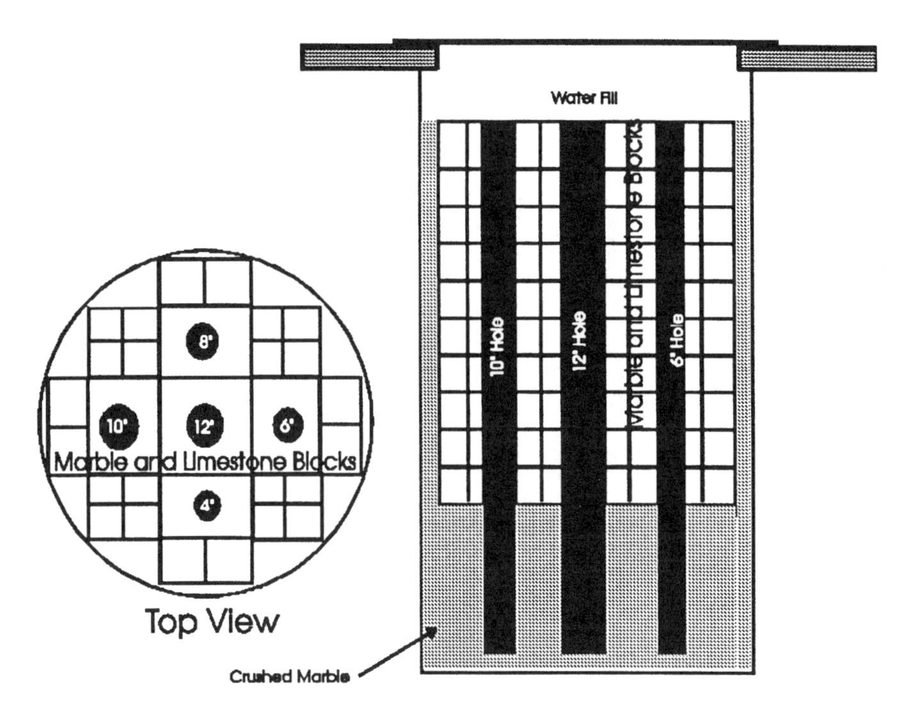

FIGURE 11.9
An acoustic calibration model.

11.4.3 Neutron Systems

Oilfield neutron systems are mostly porosity measuring systems. The experience of the contractor with the acoustic system model was borrowed by the American Petroleum Institute (API) to design and install a neutron calibration model at the University of Houston. This was done to eliminate the confusion caused by the use of many different scale units and detector responses among the various logging contractors (refer to Belknap et al, 1959). To accomplish this, a model was constructed containing three different saturated porosities. The 19% porosity Indiana limestone was arbitrarily designated to give a response of 1000 API neutron units. This allowed the petroleum industry to have neutron porosity logs with standard scales, and this allowed the comparison of any neutron porosity logs. Figure 11.10 shows a typical neutron porosity calibration model.

FIGURE 11.10
A neutron porosity calibration model.

11.4.4 Gamma Ray Systems

Somewhat the same confusion was present with the gamma ray logs used in the oilfield. A different scaling unit existed for each contractor and for each instrument type. This was resolved by the installation of a gamma ray calibration model at the University of Houston, by the API. Refer to *Introduction to Geophysical Formation Evaluation*. Because the petroleum industry did not need an absolute calibration of the gamma ray (indeed, many believed it was not possible), the model arbitrarily designated the response of the center zone of the model to be 200 API gamma ray units. This zone had been designed to have the radiation equivalent to twice that of "a typical mid-continent shale". This was successful in eliminating much of the confusion. It was not an absolute calibration, however.

The mineral industry demanded an absolute, well-calibrated log when the uranium industry grew. Because the response of every detector and every instrument was different, this demanded a universal calibration system. The U.S. Atomic Energy Commission (now part of the U.S. Department of Energy) designed and installed a wide range of calibration models at Grand Junction, CO. They also installed good secondary models at several locations in the U.S. The general form of these is shown in Figure 11.11. The instruments were all assigned conversion factors ("K"-factor) and scales in equivalent uranium oxide percentages (eU_3O_8). The method was so successful that it has been copied all over the world (i.e., Canada, Australia, India, Germany). The method has also been adapted to spectrometric gamma ray systems, airborne equipment, and potassium evaporites.

The method of the U.S. Department of Energy consisted of using graded, sedimentary uranium ore, which was in equilibrium. Measured amounts of this ore were added to the sand of concrete and used for the model. Care was taken that the heavy uranium compound did not settle before the concrete set up. The models were kept saturated with water, except for some which were deliberately left to dry. Models were made to check tool response in several grades, different hole sizes, and water- or air-filled boreholes.

A similar model for potassium evaporites could seal evaporite material in an aluminum container. The wall of the borehole should be as thin as possible. The borehole should slant slightly to ensure that the tool remains against the hole wall, as it does in the borehole.

11.4.5 Density Systems

Calibration of density systems presents different problems. The selection of materials for the model is not difficult. One must stay away from very heavy materials or materials with heavy components to avoid excessive photoelectric reactions. Because the density system uses scattered gamma rays, radioactive materials must be avoided. The material should be uniform. Granular material, if it is to be wet, must be very carefully impregnated to avoid

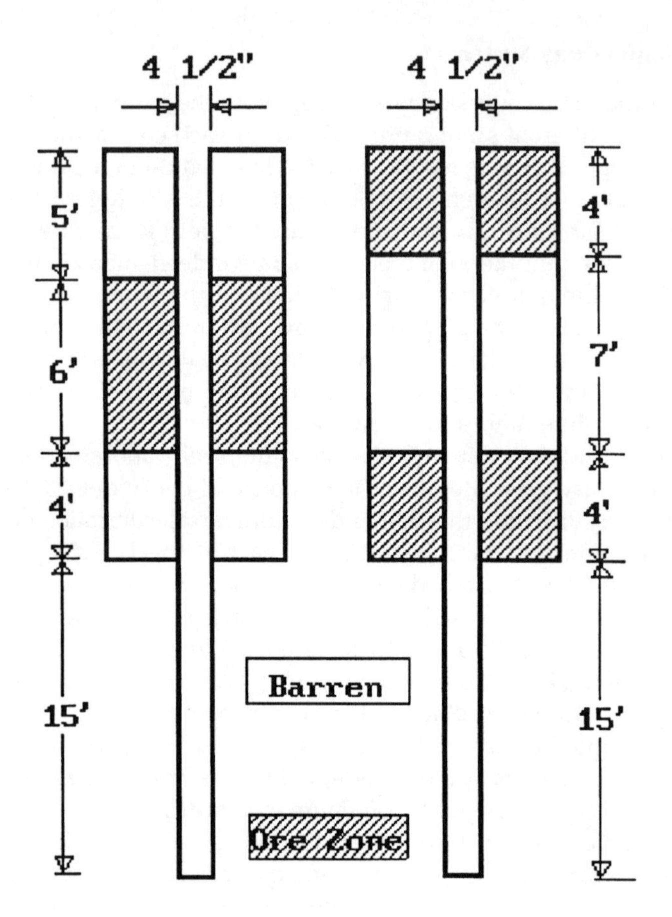

FIGURE 11.11
An example of the mineral-type calibration model for gamma ray systems.

entrapped air. Traces of surface tension lowering compounds, such as detergents, are advisable.

Density model materials can and have included water, oils, polystyrene, magnesium, aluminum, limestone, marble, dolomite, basalt, sand, glass, and coal. Basalt, sand, and glass must be checked for radioactivity. Coal tends to be nonuniform.

If the model uses granular material, such as sand or glass beads, and filled with water or oil, care must be taken to avoid entrapped air. It helps to use fairly coarse grained materials. If possible, the model should be evacuated prior to introducing the liquid. Water can have a small amount of detergent added to it. Sometimes alcohol or propylene glycol can be used or added to the water. This will lower the surface tension somewhat and also help prevent freezing. Introduce the liquid slowly. Seal the container after filling it. Any borehole casing material should be thin. Aluminum tubing makes a good casing.

Some logging contractors, whose business is primarily coal, have attempted to use blocks of coal as calibrating models. The problem is that coals are nonuniform, as far as a calibrating material is concerned. Thus, a sample taken for determining parameters may not represent the whole body of the model. A far better material to use is a block of polystyrene or polypropylene. These materials are readily available in sheet, block, and granular form. They have about the same density range as the lighter coals: bituminous, lignite, and peat.

The density of the model material is easily verified. It is merely the weight of the material divided by its volume. Commercial scales and platform balances are surprisingly accurate and are usually readily available.

Regression Analysis

Another powerful analytical method for examining geophysical data is regression analysis or curve fitting. This method examines clustered trends of data and evolves a single, probable equation that best fits the trend. The value of this method lies in the fact that our measurements are not perfect, that many parameters contributing to the measurement are not identified, and our systems always have some degree of "noise".

In regression analysis, each reading of an instrument, map, or log can be plotted on an x,y plot (or an x,y,z plot). The value of "x" may be amplitude, for example, and the value of "y" the depth. Because of the variances of each value, the cluster of plotted points will show a pure trend (caused be the desired information) and a scatter (caused by the other variables and noise).

The theoretical pure trend may follow one of a group of curves, for example, whose equations are

1. Linear: $y = a + bx$
2. Power: $y = a + x^b$
3. Logarithmic: $\log y = \log a + b \log x$
4. Exponential: $y = a\, 10^{bx}$

All four are variations of the first equation, as

$$y = a' + b'x$$

where a′ and b′ can be any function of a and b, respectively.

If a number of points are plotted on a graph paper, because of the measurement errors and variables, the points will form a scattered cluster. Examining the cluster, we may be able to determine an average or mean trend of these points. It may be informative, also, to have a quantitative value assigned to the amount of scatter (that will show the accuracy of the measurement and the degree to which we can trust it). A line drawn on the mean trend might represent the ideal value of the measurements and describe it with an equation. The values of a′ and b′ of the above general equation represent the y-intercept and the mean slope, respectively of the ideal equation. Thus, if the mean y-intercept, a′, and the mean slope, b′, can be identified, a probable mean curve can be drawn and theoretical or ideal values be assigned.

Further, the tighter the cluster of points is, the greater our confidence that the calculated line truly represents the ideal or correct case. A value representing the relative lack of scatter is called the standard deviation and it can be calculated.

Mean Values

The mean value of "x" and "y" can be obtained by,

$$\bar{x} = \sum \frac{x}{n} \tag{A.1a}$$

and

$$\bar{y} = \sum \frac{y}{n} \tag{A.1b}$$

The variance, s^2, of the values of "y" are

$$s^2 = \frac{\sum (y - \bar{y})^2}{n - 1} \tag{A.2}$$

and the standard deviation, s_y, is

$$s_y = \left(\frac{\left(\sum (y - \bar{y})^2 \right)}{(n - 1)} \right)^{1/2} \tag{A.3}$$

The slope of the curve, b, can be found

$$b = \frac{\sum (x - \bar{x})(y - \bar{y})}{\sum (x - \bar{x})^2} \tag{A.4}$$

and the y-intercept, a, is

$$a = \bar{y} - b\bar{x} \tag{A.5}$$

B

Matrix Algebra

Procedures, such as cross plotting, imaging logs, and other methods of presentation and analysis often require solving a simultaneous set of equations. In the Chapter 9, Section 9.11.1, entitled Cross Plotting, we saw a method which was originally, and primarily continues to be, solved graphically. These problems are also being solved by treating a set of simultaneous algebraic equations. This requires solving first an algebraic equation for one unknown and substituting that solution into the next equation, until all unknowns have been quantified. This can be lengthy and tedious.

The simultaneous equations can be solved by a matrix method. This also lends itself easily to computer programming. The matrix method involves assembling the unknown terms in a grid and cross operating to solve for the unknowns.

Basically, the matrix method is, as follows:

If two or more simultaneous equations are

$$a_1x + b_1y = k_1 \tag{B.1a}$$

$$a_2x + b_2y = k_2 \tag{B.1b}$$

then

$$(a_1b_2 - a_2b_1)x = k_1b_2 - k_2b_1 \tag{B.1c}$$

and

$$(a_1b_2 - a_2b_1)y = a_1k_2 - a_2k_1 \tag{B.1d}$$

This is customarily written symbolically:

$$a_1b_2 - a_2b_1 = \begin{bmatrix} a_1 & b_1 \\ a_2 & b_2 \end{bmatrix} \tag{B.3}$$

The symbol

$$\begin{bmatrix} a_1 & b_1 \\ a_2 & b_2 \end{bmatrix} = D \tag{B.4}$$

Then

$$x = \frac{\begin{bmatrix} k_1 & b_1 \\ k_2 & b_2 \end{bmatrix}}{D} \tag{B.5a}$$

and

$$y = \frac{\begin{bmatrix} a_1 & k_1 \\ a_2 & k_2 \end{bmatrix}}{D} \tag{B.5b}$$

For three unknowns,

$$D = \begin{bmatrix} a_1 & b_1 & c_1 \\ a_2 & b_2 & c_2 \\ a_3 & b_3 & c_3 \end{bmatrix} \tag{B.6}$$

and

$$x = \frac{\begin{bmatrix} k_1 & b_1 & c_1 \\ k_2 & b_2 & c_2 \\ k_3 & b_3 & c_3 \end{bmatrix}}{D_3}, \quad y = \frac{\begin{bmatrix} a_1 & k_1 & c_1 \\ a_2 & k_2 & c_2 \\ a_3 & k_3 & c_3 \end{bmatrix}}{D_3}, \quad z = \frac{\begin{bmatrix} a_1 & b_1 & k_1 \\ a_2 & b_2 & k_2 \\ a_3 & b_3 & c_3 \end{bmatrix}}{D_3} \tag{B.7}$$

the solution for three equations (i.e., for "x") is

$$x = \frac{k_1b_2c_3 - b_1c_2k_3 + c_1k_2b_3}{a_1b_2c_3 - a_2b_3c_1 + a_3b_1c_2} \tag{B.8}$$

Any number of unknowns can be handled in a similar manner. It is important to remember, however, that there *must* be the same number of different equations as the number of unknowns.

C

Bibliography and References

Albright, J.N. and Pearson, C.F., Acoustic transmissions as a tool for hydraulic fracture location, *JPE*, August 1982.

Alger, R.P. and Harrison, C.W., Improved Fresh Water Assessment in Sand Aquifers Utilizing Geophysical Well Logs, *The Log Analyst*, Vol. 30, No. 1, SPWLA, Houston, January to February 1989.

Allaud, L. and Martin, M., *Schlumberger, The History of a Technique*, John Wiley & Sons, New York, 1977.

Allen, L.S., Tittle, C.W., Mills, C.W., and Caldwell, R.L., Dual-spaced Neutron logging for porosity, *Geophysics*, Vol. 32, no. 1, pp. 60-68, 1967.

Amyx, J.W., Bass, D.M., and Whiting, R.L., *Petroleum Reservoir Engineering*, McGraw-Hill, New York, 1960.

Austin, J. and Faulkner, T., Magnetic resonance imaging log evaluates low-resistivity pay, *The American Oil and Gas Reporter*, Numar, c1990.

Bailey, R.V. and Childers, M.O., *Applied Mineral Exploration with Special Reference to Uranium*, Westview Press, Boulder, CO, 1977.

Barnes, H.L., *Geochemistry of Hydrothermal Ore Deposits*, 2nd ed., John Wiley & Sons, New York, 1979.

Belknap, W.B., Dewan, J.T., Kirkpatrick, C.V., Mott, W.E., Pearson, A.J., and Rabson, W.R., *API Calibration Facility for Nuclear Logs*, Revision of RP33, Recommended Practice for Standard Calibration and Form for Nuclear Logs, American Petroleum Instutute, Houston, TX, 1959.

Bhatnagar, A.S., *A Study on the Behaviour of Radon in Soil*, Uranium Exploration Methods, International Atomic Energy Agency, Vienna, 1973.

Bigelow, E.L., *Fundamentals of Diplog Analysis*, Dresser Atlas, Dresser Industries, Houston, TX, 1987.

Boatman, E.M., An Experiment of Some Relative Permeability — Relative Electrical Conductivity Relationships, unpublished Master's Thesis, Dept. of Pet. Eng., University of Texas, Austin, June, 1961.

Burke, J.A., Schmidt, A.W., and Campbell, R.L., Jr., The litho porosity cross plot, SPWLA, *The Log Analyst*, Vol. X, No. 6, Houston, TX, December 1969.

Cassel, B., *Vertical Seismic Profiles — An Introduction*, Western Geophysical Company, Middlesex, U.K., 1984.

Calhoun, J.C., Jr., *Fundamentals of Reservoir Engineering*, University of Oklahoma Press, Norman, 1955.

Campbell, W.M. and Martin, J.L., Displacement logging — a new exploratory tool, *J. Petroleum Technol.*, 233-239, December 1955.

Carmen, P.C., *Flow of Gases Through Porous Media*, Academic Press, New York, 1956.

Chapellier, D., *Diagraphies Appliquees A L'Hydrologie*, Lavoisier TEC & DOC, Paris, 1987.

Chapellier, D., *Well Logging in Hydrogeology*, Oxford & IBH Publishing, New Delhi, 1992.

Chaussier, J.-B. and Morer, J., *Mineral Prospecting Manual*, Elsevier, New York, 1987.

Clark, I., *Practical Geostatistics*, Applied Science Publishers, London, 1979.

Clavier, C. and Rust, D.H., Mid plot: a new lithology technique, SPWLA, *The Log Analyst*, Vol. XVII, No. 6, Houston, TX, November to December 1976.

Coates, G.R., Peveraro, R.C.A, Hardwick, A., and Roberts, D., *The Magnetic Resonance Imaging Log Characterized by Comparison with Petrophysical Properties and Laboratory Core Data*, Society of Petroleum Engineers 22723, 1991.

Coates, G.R., Vinegar, H.J., Tutunjian, P.N., and Gardner, J.S., *Restrictive Diffusion from Uniform Gradient NMR Well Logging*, Society of Petroleum Engineers 26472, 1993.

Conaway, J., Temperature logging as an aid to understanding groundwater flow in boreholes, *Proc. 2nd Int. Symp. Borehole Geophysics for Minerals, Geotechnical and Groundwater Applications, Minerals and Geotechnical Logging Society (MGLS)*, Golden CO, October 1987.

Coolidge, J.E. and Gamson, B.W., Present status of nuclear magnetic logging, Byron Jackson Division of Borg-Warner Corp., c1960.

Cork, J.M., *Radioactivity and Nuclear Physics*, D. Van Nostrand, Princeton, NJ, 1957.

Crain, E.R., *The Log Analysis Handbook*, PennWell Publishing, Tulsa, OK, 1986.

Crew, M.E. and Berkoff, E.W., *Twopit, A Different Approach to Calibration of Gamma Ray Logging Equipment*, U. S. Atomic Energy Commission, Grand Junction, CO, 1969

Czubek, J.-A., and Lenda, A., Coefficients used in the digital interpretation of gamma-ray logs, Report No. 1042/PL, Institute of Nuclear Physics, Kraków, Poland, 1979.

Czubek, J.-A, and Lenda, A., G-function in gamma-ray transport problems, Report No. 1042/PL, Institute of Nuclear Physics, Kraków, Poland, 1978.

Dakhnov, V.N., (Keller, G.V., Ed.), Geophysical well logging, *Colorado School of Mines Q.*, Vol. 57, No.2, Golden, CO, 1962.

Darcy, H., *Les Fontaines Publiques de la Ville de Dijon*, Victor Dalmont, Paris, France, 1856.

Degolyer, E., Notes on the early history of applied geophysics in the petroleum industry, *J. Soc. Petrol. Geophys.*, Division of Geophysics, AAPG, Vol. VI, No. 1, July 1935.

Dewan, J.T., *Essentials of Modern Open Hole Log Interpretation*, PennWell Books, Tulsa, OK, 1983.

Desbrandes, R., *Encyclopedia of Well Logging*, Gulf Publishing, Houston, TX, 1985.

Doll, H.G., *Introduction to Induction Logging and Application to Logging of Wells Drilled with Oil Base Mud*, Petroleum Transactions, AIME, Texas, June 1949.

Doll, H.G., *The SP Log in Shaly Sands*, AIME Paper T.P. 2912, 1949.

Doveton, J.H., *Geologic Log Analysis Using Computer Methods*, The American Association of Petroleum Geologists, Tulsa, OK, 1994.

Doveton, J.H., *Log Analysis of Subsurface Geology*, John Wiley & Sons, New York, 1986.

Dresser Atlas, *Well Logging and Interpretation Techniques*, Dresser Industries, Houston, 1982.

Dresser Atlas, *Log Interpretation Charts*, Dresser Industries, Houston, 1985.

Duda, L.E. and Pitman, J.K., *Pore Structure Analysis of Sandstones using Computer-Processed Photomicrographs*, Sandia National Laboratories, SAND82, UC-92, Albuquerque, N M, November 1982.

Edwards, A.L., *An Introduction to Linear Regression and Correlation*, W.H. Freeman, San Francisco, 1976.

Eisler, P.L., Huppert, P., and Wylie, A.W., Logging of copper in simulated boreholes by gamma spectroscopy. 1. Activation of copper by fast neutrons, *Geoexploration*, 9, pp. 181-194, 1971.

Ellis, D.V., *Well Logging for Earth Scientists*, Elsevier Science Publishing, New York, 1987.

Englehart, W.V. and Pitter, H., Über die Zusmamenhangen Zwischen Porositat, permeabilitat, und Korgrobe bei Sanden und Sandstein, *Heidel. Beitr. Petrogr.*, 2, 1951.

Fertl, W.H., *Status of Shaly Sand Evaluation*, CWLS 4th Formation Evaluation Symposium, paper 1, Calgary, canada, May 1972.

Fertl, W.H., *Gamma Ray Spectral Data Assists in Complex Formation Evaluation*, Transactions, 6th SPWLA European Formation Evaluation Symposium, London, England, 1979.

Fertl, W.H. and Wichmann, P.A., How to determine static BHT from well log data, *World Oil*, January 1977.

Frasier, D.C., Keevil, N.B., Jr., and Ward, S.H., Conductivity spectra of rocks from the Craigmont ore environment, *Geophysics*, Vol. 29, no. 5, pp. 832-847, 1964.

Friedman, G.M. and Sanders, J.E., *Principles of Sedimentology*, John Wiley & Sons, New York, 1978.

Frost, E., Fertl and W.H., *Integrated Core and Log Analysis Concepts in Shaly Clastic Reservoirs*, CWLS Meeting, Calgary, Canada, October 1979

Garrels, R.M. and Christ, C.L., Solutions, Minerals, and Equilibria, Freeman, Cooper & Company, San Francisco, 1965.

Geyer, R.L. and Myung, J.I., *The 3-D Velocity Log, A Tool for In-Situ Determination of the Elastic Moduli of Rocks*, Seismograph Service Corporation, Tulsa, OK, 1970.

Gilluly, J., Waters, A.C., and Woodford, A.O., *Principles of Geology*, Third Ed., W.H. Freeman, San Francisco, 1968.

Glossary of Terms and Expressions Used in Well Logging, Second Ed., Society of Professional Well Log Analysts, Houston, 1984.

Gondouin, M., Tixier, M.P., and Simard, G.L., An experimental study on the influence of the chemical composition of electrolytes on the SP curve, *J. Petrol. Technol.*, Feb. 1957.

Goldberg, D.E. and Dillard, C.R., *College Chemistry*, Macmillan, New York, 1974.

Green, W.R., *Computer-Aided Data Analysis*, John Wiley & Sons, New York, 1985.

Gray, H.B. and Haight, G.P., Jr., Basic Principles of Chemistry, W.A. Benjamin, New York, 1967.

Guyod, H., Temperature logging, *The Oil Weekly*, Houston, TX, 1944.

Guy, J.O., Smith, W.D.M., and Youmans, A.H., *The Sidewall Acoustic Neutron Log*, Conference paper: SPWLA Twelfth Annual Logging Symposium, May 2-5, 1971.

Hallenburg, J.K., Ed., Duray, J., Furlong, V.L.F., Kileen, P.G., and Barretto, P., *Borehole Logging for Uranium Exploration*, Technical Report Series No. 212, International Atomic Energy Agency, Vienna, Austria, 1982.

Hallenburg, J.K., *Geophysical Logging for Mineral and Engineering Applications*, PennWell Books, Tulsa, OK, 1984.

Hallenburg, J.K., Ed., *Geothermal Log Interpretation Handbook,* Society of Professional Well Log Analysts, Houston, TX, 1982.
Hallenburg, J.K., *HP41C Formation Evaluation Programs*, PennWell Books, Tulsa, OK, 1984.
Hallenburg, J.K., *Logocomp, Petroleum Formation Evaluation*, PennWell Books, Tulsa, 1985.
Hallenburg, J.K., Mineral Logging, Oil and Gas Consultants International, Inc., Tulsa, OK, 1980.
Hertzog, R.C., Laboratory and field evaluation of an inelastic neutron scattering and capture gamma ray spectrometry tool, *Soc. Petrol. Eng. J.*, 327-340, 1980.
Hewlett-Packard Owner's Handbook, HP41C, Corvallis, Oregon, 1985.
Holt, O.R., *Relating Diplogs to Practical Geology,* Dresser Atlas, Dresser Industries, Houston, TX, 1980.
Ives, D.J.G. and George, J.J., Eds., *Reference Electrodes*, Academic Press, New York, 1961.
Jackson, L.J., *Geophysical Examination of Coal Deposits*, Report no. ICTIS/TR 13, IEA Coal Research, London, 1981
Keller, G.V. and Frischknecht, F.C., *Electrical Methods in Geophysical Prospecting*, Pergamon Press, Oxford, 1966.
Keller, G.V., Electrical prospecting for oil, *Colorado School of Mines Q.*, Vol. 63, No. 2, April 1968.
Keller, G.V., Induction logging, *Colorado School of Mines Q.*, Vol. 57, No. 2, April 1962.
Konen, C.E., *Deriving Empirical Equations,* SPWLA Logging Symposium Transactions, 21st Annual Symposium, Lafayette, LA, 1980
Koerperich, E.A., Shear wave velocities determined from long and short spaced borehole acoustic devices, *JPE*, Dallas, TX, 1980.
Koerperich, E.A., Investigation of acoustic boundary waves and interfering patterns as techniques for detecting factures, *JPE*, Dallas, TX, 1978.
Lapp, R.E. and Andrews, H.L., *Nuclear Radiation Physics*, Prentice-Hall, New York, 1949.
Larinov V.V., *Borehole Radiometry,* Nedra, Moscow, 1969.
Lawson, B.L., Cook, C.F., and Owen, J.D., A Theoretical and Laboratory Evaluation of Carbon Logging. IV. Laboratory Evaluation, SPE Paper 2960, AIME, 1970
Lawrence, T.D., Continuous Carbon/Oxygen Log Interpretation Techniques, Society of Petroleum Engineers Paper 8366, AIME, 1979.
LeRoy, L.W. and LeRoy, D.D., *Subsurface Geology,* Colorado School of Mines, Golden, CO, 1977
Matthews, M.A. and Newman, K.L., *Geothermal Well Logging/Test Wells,* U.S. Department of Energy, Los Alamos Scientific Laboratory, Los Alamos, New Mexico, c1975.
McNeill, J.D., Applications of Transient Electromagnetic Techniques, Technical Note TN-7, Geonics Limited, 1745 Meyerside Drive, Missisauga, Ontario, Canada L5T 1C5, 1980.
McNeill, J.D., Hunter, J.A., and Bosnar, M., Application of a borehole magnetic susceptibility logger to shallow lithological mapping, *J. Environ. Eng. Geophys.*, Vol. 0, No. 2, 1996.
Miller, J.M. and Ostle, D., Radon Measurement in Uranium Prospecting, Uranium Exploration Methods, International Atomic Energy Agency, Vienna, 1973.
Meehan, D.N. and Vogel, E.L., *HP41C Reservoir Engineering Manual*, PennWell Books, Tulsa, OK, 1982.

Meyers, J.E., Jr., High Temperature Helium-3 Detectors, Conference paper, Institute of Electrical and Electronic Engineers, Boston, MA, October 19-21, 1966.

Miller, M.N., Paltiel, Z., Gillen, M.E., Granot, J., and Bouton, J.C., Spin Echo Magnetic Resonance Logging: Porosity and Free Fluid Index Determination, Society of Petroleum Engineers 20561, 1990.

Moran, J.H. and Kunz, K.S., Basic theory of induction logging and application to study of two-coil sondes, *Geophysics*, Vol. XXVII, No. 6, Part 1, 1962.

Myung, J.I. and Helander, D.P., Correlation of Elastic Moduli Dynamically Measured by *In-Situ* and Laboratory Techniques, 13th Annual Logging Symposium, SP-WLA, Tulsa, OK, 1972.

Myung, J.L. and Henthorne, J., Elastic Property Evaluation of Roof Rocks with 3-D Velocity Logs, Solution Mining Research Institute, Atlanta, 1971.

Mwenifumbo, C.J., Drillhole Mise-Á-La-Masse, Induced Polarization and Potential Measurements in a Zn-Pb-Cu Sulfide Deposit, Proceedings of the 1st MGLS International Symposium (held in Toronto), Geological Survey of Canada paper 85-27, Ottawa, Canada, 1983.

Mwenifumbo, C.J., Crosshole Mise-Á-La-Masse Mapping of Fracture Zones at Bell's Corner Geophysical Test Area, Ottawa, Canada, Paper M, 2nd MGLS Symposium for Minerals, Geotechnical Engineering and Groundwater Applications, Golden, CO, 1987.

Nilsson, B., *A New Borehole Radar System*, !st International Symposium in Borehole Geophysics for Mining and Geotechnical Applications (in Toronto), MGLS-SPWLA, Houston, TX 1983.

NL Baroid/NL Industries, Inc., *Manual of Drilling Fluids Technology, The History and Functions of Drilling Mud*, Vol. 1, Section 1, 1979.

Numar, Magnetic Resonance Image Logging, advertising folder, Houston, TX, c1992.

Oliver, D.W., Frost, E., and Fertl, W.H., *Carbon/Oxygen Log*, Dresser Atlas, Dresser Industries, Houston, TX, 1981.

Oliver, R.D. and Lavelle, M.J., Borehole Inspection System for Large Diameter Holes, Paper DD, Proceedings of the 2nd International Symposium on Borehole Geophysics for Minerals, Geotechnical, and Groundwater Applications, Golden, CO, 1987.

Overton, H.L. and Lipson, L.B., A Correlation of Electrical Properties of Drilling Fluids with Solid Content, AIME, 213:333-336, 1958.

Pirson, S.J., Effect of Anisotropy on Apparent Resistivity Curves, *Bull. AAPG*, Vol. 19, no. 1, 37-57, 1935.

Quantitative Mineral Exploration, Colorado School of Mines, Vol. 68, Number 1, Golden, CO, January 1973.

Random House Dictionary of the English Language, Random House, New York, 1967.

Raymer, L.L. and Biggs, W., *Matrix Characteristics Defined by Porosity Computations*, Schlumberger Well Services, c1970.

Recognition and Evaluation of Uraniferous Areas, Proceedings of a Technical Committee Meeting, International Atomic Energy Agency, Vienna, Austria, 17-21 November 1975.

Reference Data for Radio Engineers, Fifth Edition, Howard W. Sams, Indianapolis, IN, 1968.

Roy, A. and Apparao, A., Depth of investigation in direct current methods, *Geophysics*, Vol. 36, No. 5, 943-959, 1971.

Roy, A. and Dhar, R.L., Relative contribution to signal by ground elements in two-coil induction logging systems, *Geophysics*, 1969.

Recommended Practice for Determining Permeability of Porous Media, American Petroleum Institute, APR RP No. 27, Sept. 1952.
Sandberg, E., Olsson, O., and Falk, L., *Combined Interpretation of Fracture Zones in Crystalline Rock Using Single Hole, Cross Hole, and Directional Radar Data,* Proceedings of the Third International Symposium on Borehole Geophysics for Mining, Geotechnical, and Groundwater Applications (in Las Vegas, NV); MGLS-SPWLA, Houston, TX, 1989.
Schlumberger, A.G., *The Schlumberger Adventure,* Arco Publishing, New York, 1982.
Schlumberger Dipmeter Interpretation, Vol. 1, Schlumberger Limited, New York,1983.
Schlumberger Dipmeter Interpretation, Vol. 2, Schlumberger Limited, New York,1983.
Schlumberger Educational Services, *Log Interpretation Principles/Applications,* Houston, TX, 1987.
Schlumberger Well Services, Inc., *Log Interpretation Charts,* 1986.
Schlumberger Well Services, Inc., *Phasor Induction Service,* Company Brochure, Houston, TX, c1993.
Sears, W.S., *Electricity and Magnetism,* Addison-Wesley, Cambridge, MA, 1955.
Sentfle, F.E. and Hoyte, A.F., Mineral exploration and soil analysis using *in situ* neutron activation, *Nucl. Instrum. Meth.,* Vol. 42, 93-103.
Sharma, P.V., *Geophysical Methods in Geology,* Elsevier, Amsterdam, 1986.
Sheriff, R.E., *Encyclopedic Dictionary of Exploration Geophysics,* Society of Exploration Geophysics, Tulsa, OK, 1973.
Shortly, G. and Williams, D., *Physics,* Prentice-Hall, New York, 1950.
Simposio Sobre Prospeccion de Carbon, Area 1, Programacion de la Investigacion, Organiza la Escuela Técnica Superior de Ingenierosde Minas de la Universidad de Oviedo, Oviedo, 1982.
Simposio Sobre Prospeccion de Carbon, Area 2, Ejecucion de Sondreos, Organiza la Escuela Técnica Superior de Ingenierosde Minas de la Universidad de Oviedo, Oviedo, 1982.
Simposio Sobre Prospeccion de Carbon, Area 3, Testificacion, Organiza la Escuela Técnica Superior de Ingenierosde Minas de la Universidad de Oviedo, Oviedo, 1982.
Smith, O.C., *Identification and Qualitative Chemical Analysis of Minerals,* D Van Nostrand, Princeton, NJ, 1953.
Sokolnikoff, I.S. and Sokolnikoff, E.S., *Higher Mathematics for Engineers and Physicists,* McGraw-Hill, New York, 1941.
Spock, L.E., *Guide to the Study of Rocks,* Harper & Brothers, New York, 1953.
Swulus, T.M., Porosity calibration of neutron logs, SACROC unit, *J. Petrol. Technol.,* 1986.
Technical Report Series No. 158, Recommended Instrumentation for Uranium and Thorium Exploration, International Atomic Energy Agency, Vienna, Austria, 1974.
Technical Report Series No. 174, Radiometric Reporting Methods and Calibration in Uranium Exploration, International Atomic Energy Agency, Vienna, Austria, 1976.
Technical Report Series No. 208, Remote Sensing in Uranium Exploration, International Atomic Energy Agency, Vienna, Austria, 1981.
Technical Report Series No. 212, Borehole Logging for Uranium Exploration, A Manual, International Atomic Energy Agency; Vienna, Austria, 1982.
The Art of Ancient Log Analysis, Society of Professional Well Log Analysts, 1979.
Timur, A., An investigation of permeability, porosity, and residual water saturation relationships for sandstone reservoirs, *The Log Analyst,* July-August, 1968.

Tittle, C.W., *How to Compute Absorption and Backscatter of Gamma Rays*, Nuclear Chicago Corporation, Des Plaines, IL.

Tittman, J., *Geophysical Well Logging*, Academic Press, Orlando, FL, 1986.

Tixier, M.P., Evaluation of permafrost from electric log gradients, *Oil Gas J.*, June 16, 1949.

Uranium Exploration Methods, Proceedings of a Panel, International Atomic Energy Agency; Vienna, Austria, 1973.

Vennard, J.K., *Elementary Fluid Mechanics*, John Wiley & Sons, New York, 1961.

Von Seggern, D.H., *CRC Handbook of Mathematical Curves and Surfaces*, CRC Press, Boca Raton, FL, 1990.

Weast, R.C., Ed., *CRC Handbook of Chemistry and Physics*, 61st ed., CRC Press, Boca Raton, FL, 1981.

Wichmann, P.A., Hopkinson, E.C., and McWhirter, V.C., *The Carbon/Oxygen Log Measurement*, SPWLA, Dresser-Atlas Publication, 1977.

Wills, A.P., *Vector Analysis with an Introduction to Tensor Analysis*, Dover Publications, New York, 1931.

Winn, R.H., A report on the displacement log, *J. Petrol. Technol.*, February 1958.

Winsauer, W.O., Shearin, H.M., Jr., Masson, P.H., and Williams, H., 1952, Resistivity of brine saturated sands in relation to pore geometry, *Am. Assoc. Petrol. Geol. Bull.*, Vol. 36, no. 2, 253-277.

Youmans, A.H., Wilson, J.C., Lebreton, B.F., and Oshry, H.I., Lane-Wells company paper, date unknown.

Zemenek, J., Low-resistivity hydrocarbon-bearing sand reservoirs, Society of Petroleum Engineers paper 15713, Dallas, TX, 1987.

D

Symbols and Abbreviations, Subscripts, Superscripts

A	Absolute, area, atomic weight (mass) in atomic mass units
a	Apparent or partially or wholly uncorrected (subscript), tortuosity factor
AC, ac, A.C.	Alternating current
Ag	Silver, element no. 47
α	Alpha particle, proton (may also be a subscript)
Å	Ångstrom unit = 10^{-8} cm
AMU, amu	Atomic mass unit = 9.31141×10^{8} eV, 1/12 the mass of a carbon atom
~	Approximately
$\approx$, $\simeq$	Approximately equal to
atm	Atmosphere of pressure, 1 atm = 1.01325 bar = 1033.23 g/cm^2 = 760 torr
Au	Gold, element no. 79 (may also be a subscript)
API	American Petroleum Institute, unit approved by the API
B	Magnetic flux density
b	Bulk (may also be a subscript), barn (10^{-24} cm^2/nucleus)
Ba	Barium, element no. 56
BATV	Borehole acoustic televiewer
β	Beta particle, electron, ratio v/c, (may also be a subscript)
BIB	Bibliography
BHT	Bottom hole temperature (may also be a subscript)
C	Carbon, element no. 12, centigrade, celsius, conductivity (electrical), (may be a subscript or a superscript)

Ca	Calcium, element no. 20
c	Electrical conductance, core
Cal	Caliper
cc, cm^3	Cubic centimeter
CET	Cement evaluation tool
CGS, cgs	The centimeter-gram-second system of units (may also be a subscript)
CEC, C.E.C.	Cation exchange capacity
Ci	Curie
Cl	Chlorine, element no. 17
Cl	Clay (subscript)
cm	Centimeter, 1 cm = 0.01 m
corr	Corrected (subscript)
Csg, Csg	Casing (subscript)
CGS, cgs	The international System of Units, centimeter-gram-second
χ_m, chi	Magnetic susceptibility
D	Depth, detector, bit size (diameter)
d	Diameter, darcy, deep, mathematical differential, dry, diminution, rank difference
DC, D.C., dc	Direct current
Δ	Difference
diff	Difference
dol	Dolomite, $MgCO_3$ plus water of crystallization (may also be a subscript)
DOE, D.O.E.	The U.S. Department of Energy
DPT	Deep penetration version of the electromagnetic propagation tool, (EPT)
E	Voltage, volts, voltage source (may also be a subscript)
e	Effective, electron (may also be a subscript)
EM	Electromagnetic
EMU, emu	Electromagnetic cgs unit, i EMU = 1×10^9 farads
EPT	Electromagnetic propagation tool
Erg, erg	A unit of energy, 1 erg = 1 dyn-cm = 1×10^{-7} W
est	Estimate (may also be a subscript)
ESU, ESU	Electrostatic cgs unit, 1 esu = 1.11×10^{-7} farads
eV	Electron volt, 1 eV = 1.60219×10^{-12} erg

°F	Fahrenheit degrees (temperature), °F = 9(C/5) +32, (may be a subscript or a superscript)
FE	Formation evaluation
fl	Fluid, flow line (subscript)
FMI	Fullbore magnetic imager
FMs	Formation microimaging scanner
ft	Foot, 1 ft = 0.3048 m
f, form	Formation (subscript)
F_r	Formation resistivity factor
ν	Frequency in Hz (Hertz), rate of oscillation, $\nu = 1/T$
G	Geometrical factor
g, gr, GR	Gamma ray, gram, gas (may also be a subscript)
γ	Gamma ray photon, gamma ray (may also be a subscript)
H	Hydrogen, element no. 1, magnetic field strength
h	Hydrocarbon (subscript or superscript), borehole vertical distance, Planck's constant $h = 6.626176 \times 10^{-34} J/Hz$
He	Helium, element no. 2
KUT	Potassium, uranium, and thorium (usually in reference to a three-channel spectral gamma ray log
I	Electrical current
i	Invaded, invasion, invaded zone (may also be a subscript)
IL	Induction log (subscript)
I_r	Saturation index, invaded, (may also be a subscript)
ir	Irreducible (subscript)
J	A unit of energy, $1J = 10^7$ ergs = 1 Wsec
K	Potassium, element no. 19, relative permeability, Kelvins, temperature in metric absolute degrees
k	Permeability, kilo- ($\times 10^3$)
λ	Lambda, wavelength, disintegration constant, probability constant
L, l	Length, linear distance, lag, spacing
Log, log	Mathematical logarithm (base 10), a record as a function of time or distance
Ln, ln	Mathematical logarithm (base e)
LS, LS	Limestone, $CaCO_3$ + water of crystallization (subscript)

M	Mass (quantity of matter), total rock cementation exponent (superscript)
m	Cementation exponent (superscript), meter, medium, mud (may also be a subscript)
mc	Mudcake (subscript)
mf	Mud filtrate (subscript)
Mg	Magnesium, element no. 12
MKS	The meter-kilogram-second system of units
MKSA	The meter-kilogram-second-ampere system of units
ML	MicroLog (subscript)
MLL	Microlaterolog (subscript)
MCA	Multichannel analyzer
MOP	Moveable oil plot (may also be a subscript)
MOS	Moveable oil saturation (may also be a subscript)
MSFL	Microspherically focused log (subscript)
μ	Micro ($\times 10^{-6}$), micron (10^{-6} m), viscosity, magnetic permeability
-	Negative, minus (mathematical)
MWD	Measurement while drilling (may also be a subscript)
N	Number of turns, neutron (subscript)
n	Saturation exponent, neutron, number (mathematical), (may also be a subscript)
Na	Sodium, element no. 11
NI	Ampere-turns
NML	Nuclear magnetic logging
NMR	Nuclear magnetic resonance
ν	Frequency in Hz (Hertz or cycles per second)
O	Oxygen, element no. 8
o	100% water saturated, oil, degree (superscript), initial (subscript), in a vacuum
Ω	Ohm
Ωm	Ohmmeter(s), Ohms meter(s)^2 per meter
P	Pressure
p	Pipe
Pa	Protactinium, element no. 91
Pb	Lead, element no. 82
PGT	Princeton Gamma Tech
Φ	Angle of electron scatter

ϕ	Porosity (may also be a subscript)
pe	Photoelectric (subscript)
+	Positive, Plus (mathematical)
ppb	Parts per billion
ppm	Parts per million
%	Percent, 1/100
$\propto$	Proportional to
psi	Pounds per square inch
PSP	PseudoSP (SP affected by clay mineral), usually in mV, pertaining to the PSP (subscript)
ϕ	Porosity (may also be a subscript)
Q,q	Quantity, volume per unit time
qz, qtz	Quartz
R	Electrical resistivity (may also be a subscript)
r	Electrical resistance, radius, radial distance, residual, resistivity (may also be a subscript), relative (subscript)
Ra	Radium, element no. 88
Ref	Reference, references
ρ	Density (may also be a subscript)
Rn	Radon, element no. 86
RTC	Resistivity through casing
S	Saturation, sulfur, source
s	Saturated, surrounding, sand (subscript)
s, sec, Sec	Second of time
SI	International system of units (cgs)
SH, sh	Shale (subscript)
Si	Silicon, element no. 14
σ	Micro cross section
Σ	Sum, mathematical summation, thermal neutron macro cross section
SP	Spontaneous (self) potential, usually in millivolts, pertaining to the SP (superscript)
SSP, ssp	Static SP value, usually in mV, pertaining to the SSP (superscript)
Sym, sym	Symbol
T	Time
t	True, unit (interval) travel time, circulation time, total, tortuosity

τ	Photoelectric cross section
TD	Total depth
Th	Thorium
Tl	Thallium
Θ	Photon scattering angle
θ	Angle
V	Volume
v	Velocity
VS	Volumetric scanning
VSP	Vertical seismic profiling
w	Water, wet
W	Weight, watt
WC	Water cut
WGR	Water/gas ratio
WOR	Water/oil ratio
λ	Wavelength, in cm, m, or Å
xo	Pertaining to the invaded zone
χ	Chi, the magnetic susceptibility
X-rays	A type of electromagnetic radiation of higher frequency than visible light but lower than gamma radiation, typically caused by the interaction of electrons with the K, L, M orbital electrons
Z	Atomic number
Z/A	Ratio of atomic number/atomic mass ratio

E

Definitions

Term	Definition
Bottom hole temperature (BHT)	The temperature of the borehole at the maximum depth of the borehole. This was originally determined with a maximum reading thermometer on the logging tools and originally assumed to be the maximum temperature and the formation temperature at that level.
Coulomb	The quantity of electricity transported in 1 sec by a current of 1 amp.
Diamagnetic materials	Materials within which an externally applied magnetic field is slightly reduced because of alteration of the atomic electron orbits produced by the field. The permeability of dielectric materials is slightly less than that of empty space. A consequence of the Lenz law of induction.
Dipole	A unit electrically charged or magnetic particle having a single + and a single – pole.
Dipole moment, m	The torque experienced on a dipole when at right angles to a uniform, unit electrical or magnetic field.
Faraday	The product of the Avagadro number, N_A, and the elementary charge, e. F = 96,489 ±2 coulombs per molecule.
Geophone	A specially adapted microphone designed to detect low frequency waves in the earth.
Kelvin	1/273.16 of the thermodynamic temperature of the triple point of water.
Lenz law	When an electromotive force is induced in a conductor by any change in the relation between the conductor and the magnetic field, the direction of the electromotive force is such as to produce a current whose magnetic field will oppose the change.
Magnetic	Possessing or capable of possessing a magnetic field.

Noble gas	Gases with a complete outer electron orbit, thus, inert.
Noble metal	Inert metals, such as gold.
Overburden	The earth layers above the subject deposit or formation.
Paramagnetic materials	Those within which an applied magnetic field is slightly increased by the alignment of electron orbits. The slight diamagnetic effect in materials having magnetic dipole moments is overshadowed by this paramagnetic alignment. As the temperature increases this paramagnetism disappears, leaving only diamagnetism. The permeability of paramagnetic materials is slightly greater than that of empty space.
Pixel	A unit area of a computer graphic picture or drawing.
Spacing	The distance between two objects (i.e., electrodes, coils), usually from center to center.
Specific	The English language name for an extensive physical quantity, meaning "divided by the mass".
Toroid	A coil of wire, usually cylindrical, used to create or detect a magnetic field.
Vertical Seismic Profile (VSP)	A seismic method combining surface sources with multiple downhole detectors.

F

Greek Alphabet

Name	Greek Upper Case	Lower Case	English Equivalent
Alpha	Α	α	ä
Beta	Β	β	b
Gamma	Γ	γ	g
Delta	Δ	δ	d
Epsilon	Ε	ε	e
Zeta	Ζ	ζ	z
Eta	Η	η	å
Theta	Θ	θ	th
Iota	Ι	ι	i
Kappa	Κ	κ	k
Lambda	Λ	λ	l
Mu	Μ	μ	m
Nu	Ν	ν	n
Xi	Ξ	ξ	ks
Omicron	Ο	ο	o
Pi	Π	π	p
Rho	Ρ	ρ	r
Sigma	Σ	σ	s
Tau	Τ	τ	t
Upsilon	Υ	υ	ü, öö
Phi	Φ	φ	f
Chi	Χ	χ	h
Psi	Ψ	ψ	ps
Omega	Ω	ω	o

Index

D

E

F

K

L

M

N

O

P

T

U

V